Beyond Inheritance

Beyond Inheritance

*Our Ever-Mutating Cells and
a New Understanding of Health*

ROXANNE KHAMSI

RIVERHEAD BOOKS

NEW YORK

2026

RIVERHEAD BOOKS
An imprint of Penguin Random House LLC
1745 Broadway, New York, NY 10019
penguinrandomhouse.com

Portions of chapter 1 first appeared, in different form,
as "The Darwin Treatment" in *Wired* (2019).

Portions of the introduction and chapter 5 first appeared,
in significantly different form, as "A Change of Mind"
in *MIT Technology Review* (2021).

Book design by Alexis Sulaimani

LIBRARY OF CONGRESS CONTROL NUMBER: 2025041039

ISBN 9780593541913 (hardcover)
ISBN 9780593541937 (ebook)

Printed in the United States of America
1st Printing

The authorized representative in the EU for product safety and compliance is
Penguin Random House Ireland, Morrison Chambers, 32 Nassau Street, Dublin
D02 YH68, Ireland, https://eu-contact.penguin.ie.

For Kian

CONTENTS

Introduction

Before your life ever got complicated, before you needed the warmth of an embrace or had to finish your homework, before cooking and cleaning the house and long to-do lists, before all of that, for about one day after the sperm and egg that made you combined, you existed as a single cell. At the end of this momentous first day, you divided into two cells, and then those divided again rapidly to become four. Several days later, you were a cluster of sixteen cells or more. Scientists call this the morula stage, which derives from the Latin *morum*, "mulberry." You graduated from this berrylike existence to become a mass of more than one hundred cells with a hollow center around day five. And soon after, you accomplished one of your first major feats: nestling into the inner wall of the womb in which you grew. From there on out, over the years, you kept getting bigger and your cells kept dividing. Ultimately, your body developed into a walking conglomeration of a whopping thirty trillion to forty trillion cells. That number does not even include the additional thirty-nine trillion or so tiny bacterial cells in your gut microbiome.

This multicellular existence is something people often overlook. We tend to think of ourselves as one entity—one body moving about in the world through space and time. But the mind-bending truth is that each of us is a bundle of trillions of cells working in concert to do seemingly simple things like lift a spoon to our mouth or sing a song. Here we are, ambling about as a multitude of microscopic cells.

Multicellular life took its sweet time to appear on our planet. Before it did, there were plenty of single-celled organisms. They are believed to include LUCA, which is shorthand for the last universal common ancestor. Some scientists assert that this theorized organism emerged several billion years ago and is the one from which all creatures living today descended. However, it's important to note that the transition from single-celled life to multicellularity did not just happen once on Earth. Multicellular organisms may have independently arisen more than ten times from unicellular life.

The rise of multicellularity is often told as a story about cells coming together to cooperate. When the German naturalist Ernst Haeckel observed tiny, microscopic choanoflagellates more than 150 years ago, he believed them to be the unicellular ancestors of sponges. Haeckel sought to know how the transition to multicellularity might have come to pass. In 1874, he published his hunch: He proposed that when cells of the same species congregated, they would sometimes form colonies that eventually became animals. It was a new way of thinking about how higher-developed species like ours came to be.

It's understandable if you are tempted to envision the trillions of cells in a body as coexisting in constant harmony. But the truth is that things aren't always copacetic. Cells in our bodies sometimes live in conflict. And the small—but sometimes consequential—genetic changes known as mutations that crop up in cells over time can contribute to the internal clashes. Take cancer, for example. In that disease, a rogue cell that has acquired genetic mutations can take on a life of its own, casting aside the interests of the being in which it cohabitates and hoovering up more than its fair share of resources. The malignant cell can multiply, and eventually its duplicates can decamp to other parts of the body. This horrible process, called metastasis, makes the cancer very difficult to treat.

Cancer cells are so adept at staking out new territory that they can even take over a totally different body from the one in which they originate. One tragic case from the early 1960s demonstrates this bizarre reach. In an experimental procedure, doctors intentionally transplanted a piece of skin cancer tissue measuring half a centimeter in diameter from a fifty-year-old patient into her eighty-year-old mother. It was hoped that the mother would start producing antibodies against the tumor that could be extracted and used to treat her daughter. Sadly, the daughter died from a perforated bowel the day after the transplantation took place. And within fifteen months the cancer had spread throughout the mother's body and claimed her life, too. Then there's a case in which a surgeon accidentally injured his left hand while operating on a thirty-two-year-old man in Germany to remove a malignant abdominal growth. Five months later, the surgeon had a lump in his left palm the size of a ping-pong ball. Tests revealed that it was a tumor genetically identical to the one in the patient he had operated on. The cells from the patient had apparently entered the surgeon's body during the fateful operation and tried to stage a corporeal hijacking.

It's not just cancer cells that have a propensity to multiply and edge out their cellular counterparts. As you'll read in this book, all sorts of cells have a knack for rising up within the crowd. Sometimes it's immune cells, for example, and other times it's red blood cells. There are even instances in which liver cells and brain cells outdo their neighbors. As such subsets of cells crescendo in numbers, they can have a huge influence on our health.

The key to these cells' ability to surpass others—and have big effects on our well-being—is genetic mutation.

■ ■ ■

The concept of mutation emerged long before scientists figured out the structure of DNA. The word traces back to an experiment that began in the late 1800s in a meadow near the Dutch town of Hilversum. A plot of untended land there had become overgrown with evening primrose, a plant with delicate yellow flowers that blossom at night, releasing a sweet, lemony fragrance. The botanist Hugo de Vries stumbled upon the primrose patch and determined it had escaped from a nearby garden. But that's not all he noticed. De Vries saw that some of the primrose had produced markedly different versions. So he took some seeds from the patch and over the ensuing years planted fifty thousand of them, ultimately identifying eight hundred variants that spontaneously emerged. Some had rounded leaves rather than narrow ones, while others possessed giant flowers. Some had red veins in their leaves; some were dwarfed. He considered these different species. Ultimately de Vries gave this phenomenon of biological change a name, *mutation*, in his 1901 book, *Die Mutationstheorie* (*The Mutation Theory*).

In the decades that followed, and even up until today, people have mostly focused on the role of mutation in shaping entirely new species or in occasionally causing an inherited disorder. But there is a whole world of mutation happening under the surface of our skin—and inside other creatures as well—that most of us have simply ignored.

One of the first inklings that the cells within a single organism might possess genetic differences came from the study of sea urchins more than a century ago. The German zoologist Theodor Boveri observed that sea urchin embryos with an unbalanced number of inner genetic bundles called chromosomes would develop abnormally. Boveri had the foresight to think that these anomalous cells could have real health consequences. In 1914, the same year that World War I commenced, he published a book in which he speculated that cells with chromosomal irregularities might be the root cause of malignant tumors.

Decades later, a cancer researcher in Philadelphia named Peter Nowell took this concept even further. Nowell had noticed that severely ill leukemia patients typically had more chromosomal mutations than those with less severe manifestations of the illness. This insight ignited his curiosity. In 1976, he proposed that cancers contain cells that can accrue different, additional mutations and end up in competition with *one another*.

But various findings soon eclipsed the ongoing research into mutations in individual cells. Scientists were making giant strides in DNA sequencing and using this to identify the common genetic elements that all humans share.

One big moment came on the morning of June 26, 2000. On that day, US President Bill Clinton walked into the East Room of the White House flanked by two top scientists as the brassy, patriotic tune "Hail to the Chief" played in the background. After the crowd had seated itself, Clinton took the podium. "We are here to celebrate the completion of the first survey of the entire human genome," Clinton explained. And then, with a slight lifting of his brow, he added, "Without a doubt, this is the most important, most wondrous map ever produced by humankind." The event marked a watershed moment in the history of science.

Researchers had achieved a remarkable feat. From a handful of people, they had extracted and analyzed the precious code of DNA that provides the instructions for life. The general layout of the human genome they pieced together contains roughly three billion molecular building blocks known as nucleotides, which are represented by the letters A, C, T, and G. With this rough sketch in hand, scientists were starting to get a better sense of how many genes the entire code contained. Around the time of the announcement, I was a college student studying genetics, and it seemed that no lecture in my biology courses could be complete without the professors gleefully reminding us of this giant milestone. By February 2001, the two teams working on sequencing the human genome published their official drafts a day apart from each other. And just a couple of years later, the Human Genome Project unveiled its final genome map, which covered an impressive 92 percent of all three billion nucleotides in the total sequence.

Geneticists never stopped aiming for completion. Finally, in 2023—two decades after the draft sequences were published—they formally released an end-to-end, gapless sequence of the human genome. This included the Y chromosome, whose tricky repeated bits had made it tough to decipher. The efforts to map the genome revealed along the way that humans possess at least twenty thousand genes in their DNA.

It's undeniable that researchers have made tremendous progress toward charting the genome. That said, we need a radical shift in how we conceptualize DNA and the nature of genetic disease. Let's begin with the fact that there is not just one human genome. Our genomes are around 99.5 percent similar to one another's. While that might not sound like a lot of divergence, it is. For comparison, by some measures the human genome is about 98.4 percent identical to that of the chimpanzee. The reality is that each of us inherits a slightly different version of the human DNA sequence. This much is generally known by most people.

However, what goes largely overlooked is that a multitude of genomes exist within one individual. We each start out as a single fertilized egg, but as our cells divide and mature during embryonic development and all the way until we take our last breath, they pick up genetic changes. To explain and predict the shifting patterns of cellular winners and losers within our bodies, we need a revolution in our understanding of mutation.

. . .

Traditionally, a great deal of genetics research has focused on maladies that are passed down from one generation to the next. A major breakthrough came in the late 1970s when a gene linked to the blood disorders sickle cell disease and beta-thalassemia was the first to be identified as a culprit in inherited illness. Over time, the number of hereditary diseases identified by scientists skyrocketed. Now, more than seven thousand inherited ailments have been pinned to changes in individual genes.

Standard methods for identifying mutations analyze bulk tissue but cannot discern the DNA from individual cells. You might say, perhaps, that it can see the forest but not the trees. These sequencing limitations don't present as much of a problem for studying inherited mutations. When a mutation is passed down to us from our parents, it exists in every single cell in our body. It is carried from the sperm or the egg and is present in the embryo and all cells in our body thereafter. You don't need to tease apart the DNA of different cells to find it.

Scientists only recently acquired the tools to show that cells within a single person can have diverging genetic instructions. The research community finally has a way to document the vast genetic mosaic that arises in our multicellular bodies. In the last decade, we have amassed more information about the genetic changes that accrue in an individual than we had in the previous half century.

Advances in DNA sequencing—including greater automation, higher sensitivity, and better computational tools to crunch the data obtained—have made it possible to document the mutations that arise in tissue. It has become possible to analyze bulk samples to capture the array of variations within large numbers of cells, and to do so quickly. But there have also been strides in understanding mutation at a more granular level: In 2009, scientists announced a new method for sequencing the genetic material of *individual* mammalian cells. With this major breakthrough, we could truly start to see the mutational mosaic within us. Further methods devised in subsequent years made doing so easier and cheaper. Initially, studies could separately sequence only a few cells at a time. A huge improvement came when researchers figured out a way to put cells into droplets or tiny wells in a laboratory dish so that they could analyze them individually and at a large scale. This has brought the costs of single-cell sequencing down to less than a dollar per cell.

Single-cell sequencing, which originated in a select few academic laboratories, has now reached the hands of clinical research groups around the world, from those that investigate cancer to others that study immunity and everything in between. It is also providing exquisite detail about the mosaic of cells that exist in early development. In 2018, one of the top research journals in the world, *Science* magazine, named single-cell sequencing of animal embryos the Breakthrough of the Year. All of this has shaken up the status quo of genetics. Biologists and doctors can no longer treat the cells within a body as genetic carbon copies of one another.

However, the revelation that there is mutational diversity across each body hasn't yet caught on widely among the public. When many folks contemplate DNA alterations that happen in an individual over time, they often think about shifts in epigenetic marks. These are chemical tags that sit on top of our genetic material (*epi* means "over" or "upon" in Greek). Epigenetic marks help to switch genes on or keep them suppressed. And they aren't static: They can disappear or appear over the course of a lifetime. Cigarette smoking, weight changes, and stress have all been linked to epigenetic changes. Those findings have garnered big headlines and captured the public imagination. But less attention has been paid to the fact that, beyond the changes to these chemical markers on top of DNA, during a lifetime there are actual changes occurring in the code itself. Very few people appreciate that we pick up countless mutations *within* our DNA as we age.

The idea of genetically distinct cells existing within a single human is unsurprising once you understand all the different types of mutations that can happen. A dizzying range of DNA anomalies can take place in cells to result in a mosaic body. Perhaps the most straightforward kind to conceptualize is when there are extra or missing chromosomes or other large-scale structural anomalies of the genome. There can also be biological typos affecting the "letters"—or nucleotides—in the sequence. Errors can arise even when the cell tries to repair damage to its DNA. In other instances, extra sequence repeats occur in the genome, much like an accidental copy-paste in a text document. Sometimes whole genes are duplicated, also causing trouble. Then there's the process of recombination, in which portions of DNA are swapped between strands during cell replication, fusing to create a "Frankenstein" gene with major effects.

Many of the mutations we rack up occur as a natural consequence of the inner wear and tear happening in our cells. However, they can also be accelerated by our environment and behaviors. Enjoy a cigarette from time to time? Smoking just fifteen of them might cause a mutation that becomes permanent in cells' DNA, according to one calculation. Love the beach? Every cell in sun-exposed skin picks up around one new mutation every day of life.

When it comes to mutations acquired as part of development and aging, the numbers can be staggering. Putting aside the many bacterial cells in our gut microbiome, each adult human body consists of upwards of thirty trillion human cells that all originated from a single fertilized egg cell. It took a lot of replication cycles for the cells to divide from one to thirty trillion, and each time that a cell split to form two daughter cells, it had to copy its DNA. That momentous task of copying the approximately three billion chemical "letters" of the genome is a delicate process during which errors can creep in. And the opportunity for new mutations is always there so long as we live and breathe. Our bodies are never done replenishing their tissues. Everywhere from our blood to our skin to our stomach lining requires a constant replenishment of cells for upkeep. With all this churn, mistakes in our genetic code are bound to be made.

Another way that DNA can get jumbled is when the cellular machinery that normally protects cells against invaders backfires. One example is a family of enzymes known as APOBECs. These enzymes can introduce errors in the genetic code of viruses that break into cells. But sometimes APOBECs can turn on an organism's own DNA with disastrous results. Mouse experiments have found that at least one subfamily of these enzymes can enable colon and liver tumors, and they have been implicated as a source of mutation in more than twenty human cancers.

The goal of this book is to pull back the curtain and show you that your DNA is dynamic and endlessly mutating rather than a set of instructions that remains static in your cells. This deeper understanding of the changes that accumulate within cells' genomes gives us a whole new way of thinking about our bodies. If we want to fully grasp how our genetics relates to our health, and how to use this information to fight illness, then we have to face up to the genetic variation that emerges within us over time. The public isn't typically aware of this diversification, though scientists have been offering up data illustrating it for decades. As geneticists Jan Vijg and Xiao Dong have noted, "It is clear that previous ideas of a stable genome have now been replaced by a much more dynamic view of considerable cell-to-cell variation in genome structure and function."

I want to change how you think about genetic mutations. They are not blips in the genome that happen only once in a blue moon and matter only when they are passed from one generation to the next. They are constantly happening within your body, and they can have real consequences for your personal well-being. By some estimates, you acquire trillions of new mutations a day. These arise, at varying rates, across all your tissues. Granted, many of the mutations take place in cells that subsequently die, and are a dead end. But others crop up in cells that stick around, and are carried over during regular cell division that replenishes tissues. Because of this propagation, mutations can accumulate over time within the body. Some of the most dramatic examples come from centenarians. A single white blood cell from a one-hundred-year-old can contain more than three thousand mutations, and perhaps other changes in its genetic sequence that modern technology cannot yet detect.

■ ■ ■

This book begins with our newfound understanding of genetic errors in cancer. Decades ago, scientists believed that cancers were driven by a couple of genetic errors. Modern sequencing tools have underscored just how prescient Peter Nowell was to imagine that the genetic variability within a patient's cancer increases over time. The true complexity of tumors began to reveal itself after technological improvements permitted swift sequencing of bulk samples and readouts from single cells. Now, scientists see cancers as rife with genetic change. By some estimates, an advanced tumor might harbor thousands upon thousands of mutations. And a few scientists even speculate that a "big bang" of certain kinds of genetic changes positions these malignancies to grow. Some mutations are particularly worrisome, such as the ones that empower tumor cells to sidestep the drugs meant to eradicate them.

Our new appreciation of the seemingly boundless mutations that can arise in a cancer is creating a seismic shift in how we treat the disease. By sequencing the myriad genetic changes in a tumor, we can figure out which mutations fuel its growth, and design drugs to strike these targets. This treatment approach is now possible in part because we see that a tumor is a mosaic of cells with both common *and distinct* mutations. But it's not just in cancer where this perception matters. The same insight applies to countless diseases in tissues all across the body. By letting go of the antiquated idea that every cell has the same exact DNA and embracing the messier reality that each of our cells has a slightly different genetic code, we can usher in a whole new era of medicine.

In truth, it's been more than 150 years since scientists began advocating that the cells in a human body are not all alike and do not always cooperate. As far back as the time of Charles Darwin, biologists began to propose that the forces of evolution—a concept describing the variation that makes natural selection and the emergence of new types of organisms possible—operate under the skin the same way that they operate at the species level. Their message hasn't always been heard, but the second chapter of this book traces the long trajectory of this idea.

The more we learn about the populations of cells that evolve within us, the more we see that cells' ability to mutate isn't always a bad thing. In fact, it has a vital physiological role in keeping us alive. Chapter 3 describes one of the most important functions of mutation, which is to equip us to fend off invading microbes. My hope is to persuade you that mutation is not always the villain in the story of our health—sometimes it is the hero.

The role of mutation was first appreciated in cancer and immunity, but scientists then uncovered its influence in blood conditions, including those that are life-threatening. This is the focus of chapter 4. It also explores, for example, research indicating that almost half of men over the age of seventy have lost their entire Y chromosome in many of their circulating cells. Some experts believe this loss of the Y chromosome is just the tip of the iceberg—easily visible evidence of murkier and perhaps more threatening changes that are also associated with aging. Thanks to advanced genetic sequencing, we can track the rise of populations of mutant cells in the blood not just as we age but also in nearly real time.

The list of diseases linked to spontaneous mutations keeps growing with each passing year, thanks in part to improvements in DNA testing. As that list grows, it challenges the common thinking that genetic illnesses are always inherited. People have been accustomed to hearing about genetic disorders, such as hemophilia, cystic fibrosis, and sickle cell disease, that are passed from one generation to the next. But, as you will learn in chapter 5, genetic disease is something that can also emerge within us anytime from after conception until our twilight years.

New mutations can also coincidentally *correct* inherited disorders. Since the 1990s, scientists have uncovered evidence of spontaneous genetic changes in patients that have rescued malfunctioning muscle fibers, reversed immune deficiencies, fixed dangerous skin disorders, and overcome grave blood conditions. There is a growing number of cases identified in which these genetic autocorrections have saved people's lives. As chapter 6 explains, even in people without any known inherited disease, mutation can have a beneficial effect, such as helping the liver cope with excess calories. Inspired by these kinds of findings, scientists are testing whether future therapies can give cells with beneficial mutations a leg up in the body.

There's also more attention being paid to the new mutations that crop up over time in our reproductive cells. These genetic anomalies don't typically affect our own health, but in rare cases they can be transmitted to our children. Our understanding of inherited diseases is being updated because we now appreciate that sometimes they are not passed down through many generations but instead originate with a random genetic event in parental gametes. Chapter 7 shines a light on this revelation, as well as on the study of how reproductive cells compete against one another and on efforts to screen against the transmission of new mutations in sperm so that those genetic changes are not passed on.

Of course, it's not just our own cells that mutate and grow in number inside our bodies. Genetic changes happen in the microbial passengers within us, as detailed in chapter 8. These passengers include the bacteria that help make up what's known as our gut microbiome. There are trillions of them living in our digestive tract, and their activity might even influence our waistline. The viruses that infect us can also mutate rapidly once they establish themselves. The genetic changes they pick up while replicating within an individual can determine whether that person survives their illness or succumbs to a strain that evolved within them. If a mutated virus spreads to others, it can intensify an outbreak of disease in the local community and—as we have seen with AIDS and Covid—in some instances alter the trajectory of a global pandemic.

The final chapter of this book takes a hard look at how all the mutations accumulating within us through the years shape how we age. For decades, researchers have wondered if spontaneous DNA errors can affect the vitality of all our tissues toward the end of our lives, and our risk of frailty. In scientists' search for clues about the relationship between aging and mutation, they have analyzed the DNA of organisms ranging from mice to dogs to more exotic animals such as lions, lemurs, and giraffes. Some have even ventured near the northernmost point of the United States to study the longest-lived mammal on Earth. These investigations have yielded lots of ideas about how mutation might accelerate our physical and mental degeneration over time, and what—if anything—we can do about it.

■ ■ ■

Consider this stark possibility: Even the very cells you are using to process the words of this book might be mutating.

For a long time, the brain was seen as a static organ, but that's no longer the case. One of the scientists at the forefront of this area of discovery is Christopher Walsh, a neuroscientist at Boston Children's Hospital. He and his team have been tabulating distinct mutations in carefully parsed individual brain cells. As part of one experiment, post-doctoral members of Walsh's lab and from Peter Park's lab at Harvard

Medical School helped analyze thirty-six single neurons from the preserved brains of three people who had tragically died in accidents. Although it is relatively easy for scientists to sequence the DNA of single cells these days, at the time of the experiment more than a decade ago, thirty-six was a huge accomplishment. The analysis revealed a stunning number of mutations that had piled up in the subjects' brain cells over the years. By the team's calculations, each neuron contained around fifteen hundred unique mutations. When describing the findings in *The Atlantic*, science journalist Ed Yong wrote that "each neuron really is a beautiful and unique snowflake."

Not only were there a whole bunch of mutations in the neurons, the location of these blips in DNA hinted at how they arise. Many of the errors occurred in genes that are turned on and used by a cell to carry out its normal function. When neurons unwind their genomes a bit to allow a gene to be active, they seem to provide an opportunity for damage to occur there. As a summary of the published findings exclaimed, "Neurons would seem to be their own worst enemy."

Walsh's team came out with another big study not too long after. They had sequenced brain cells from fifteen people between the ages of four months and eighty-two years, as well as nine people with disorders linked to premature aging. The data indicated that DNA changes accumulate "slowly but inexorably with age in the normal human brain." Notably, the results suggested that not everyone started on an even playing field: People born with faulty DNA repair genes were prone to acquiring brain mutations more quickly. According to the estimates of Walsh and his team, human neurons might rack up around twenty new mutations per year for life. Meanwhile, British researchers calculated that the average adult has approximately one hundred thousand to one million brain cells harboring mutations in genes linked to neurodegeneration. Their study had too small a sample size to determine whether those blips were to blame for any cases of disease, but the scale of mutation they identified is impressive.

My hope is that this book will open your eyes to the landscape of immense and ongoing genetic diversity that exists within each human body. You will recognize that you are a slightly different genetic version of yourself today from yesterday, and will be different yet again tomorrow. This constant genetic metamorphosis occurs within many organisms—from fruit flies to mice to whales. It even takes place in plants. If you happen to have a tree in your sight line right now, look up at its crown. The leaves on the swaying tips of its branches extending toward the sky possess genetic variations that are different from those in its trunk and its roots in the ground.

The living world around us is buzzing with mutation. The wellspring of variation within each multicellular organism, including each human, cannot be ignored any longer. Nor can we overlook its influence on our individual health. Many doctors and researchers now believe that mapping the universe of mutations inside us will revolutionize medical treatments. Here is the story of this awakening.

1.

Turning Cancer Against Itself

"Every one admits that the body consists of a multitude of organic units, all of which possess their own
proper attributes, and are to a certain extent independent of all others."

CHARLES DARWIN

n late 2005, Robert Butler moved to Tampa, Florida, ready to start a
new chapter in his life. As a former oil-exploration engineer from England whose work had taken him around the world, he was happy to
finally retire to the Sunshine State, where his sons had both settled. But
only a couple of years afterward, things took an unanticipated turn: He
was diagnosed with prostate cancer. Butler suffered through various
unpleasant treatments—blasts of radiation and rounds of medications
such as Lupron, which lowers testosterone levels and left him feeling
physically weak. Those traditional therapies didn't work; his prostate
tumor had reached stage 4, advanced cancer, within seven years. Butler
even tried a newly approved treatment called immunotherapy that involved having cells from his blood sent by courier to a facility outside
Atlanta, where they were mixed with a molecule that activates immune
cells, and then returned to Florida to be infused back into him. The

process was expensive—its sticker price can be as high as $120,000—but it also failed to rid him of his cancer.

In the summer of 2014, Butler and his wife showed up at his oncologist's office at the Moffitt Cancer Center, a Tampa hospital complex nestled on a lush green campus where one can occasionally encounter roosters pecking the grass outside. They braced for what would come next; they had heard about invasive treatments like radioactive seed implants. But the doctor told them about something else. He explained that a radiologist on staff had a radical new experiment and asked if Butler wanted to participate. If he signed up for the trial, Butler would take a powerful and exceedingly expensive drug called Zytiga, but not in the standard scorched-earth fashion that aims to obliterate every possible cancer cell in the body. Instead, he would receive only as much Zytiga as was necessary to stop the cancer from growing. The idea was far-out and counterintuitive. His last best shot at escaping death from his cancer was to give up on curing it.

Butler approached the study with cautious optimism. By then he had reached his seventies, and both his hair and his close-trimmed beard had turned gray. Knowing the modified Zytiga regimen wasn't designed to rid him of cancer left Butler with a burning question about how the doctors would measure the success of their new treatment approach. He asked, "How do we know this stuff is working?" Ultimately, one of his doctors replied, "Well, you won't be dead."

The cancer treatment odyssey that Butler was about to embark on was inspired largely by another journey, one that began two days after Christmas in 1831, when a twenty-two-year-old Charles Darwin set sail from Plymouth, England, onboard a ship known as the HMS *Beagle*. Over the next five years, the ship covered nearly forty thousand miles, with an often-nauseous Darwin sleeping in a hammock hung over the small table in the main cabin. "The misery I endured from sea-sickness is far beyond what I ever guessed at," he wrote in his notes.

By 1835, the ship had reached a string of volcanic islands known as the Galápagos, six hundred miles off the coast of Ecuador. In the five weeks the crew spent there, Darwin busily studied the local fauna. He filled pages in his notebooks with descriptions of the animals and plants on each island. As the HMS *Beagle* set sail, eventually making its way home to England, Darwin had the beginnings of what would become his theory of evolution. He would come to see evidence for this theory in the diversity of Galápagos finches he had encountered on his voyage. There was, for example, the large cactus finch, a black bird that weighs as much as a silver dollar, has a long, pointed beak, and, as its name suggests, snacks on cacti. Elsewhere in the archipelago there were ground finches, which had shorter beaks. The smallest of the finches in the Galápagos were the warbler finches, which boast yellow or gray feathers and weigh only slightly more than a quarter. Their beaks are thinner and more pointed than those of both the ground finches and large cactus finches—and this is where Darwin's key observations started to manifest. He noticed that the beaks of the various finches were finely tuned to help each species of bird survive on the food available in its local habitat. The slender warbler finch beaks were well adapted to help them catch insects, for example. Ultimately, he would arrive at the idea of natural selection. It would later be popularized as a concept encapsulated by the phrase "survival of the fittest"—and it sent shock waves through England and the rest of the world.

More than a century later, Darwin's writings would inspire the new cancer treatment approach at the Moffitt Cancer Center. Bob Gatenby, who developed the approach, hadn't given the concept of evolution much thought early in his career. During those days, he had been too busy working as a radiologist on the bloody front lines of the war on cancer. In the mid-1980s, he secured a job at the Fox Chase Cancer Center in Philadelphia. At that hospital and others around the country, clinical trials were putting breast cancer patients through an extreme treatment: a combination of a potentially lethal dose of chemotherapy followed by a bone marrow transplant. The treatment was harrowing. The women had diarrhea and nausea, and some had so much lung damage they had difficulty breathing. Others experienced liver damage and weakened immune systems that left them vulnerable to serious infections.

As a radiologist, Gatenby's job was to interpret X-rays and other scans of the patients, and he saw the treatment failing. Out of more than thirty thousand women with breast cancer in the US who underwent the procedure between 1985 and 1998, as many as 15 percent died from the treatment itself. "What happened was these women suffered horribly, and they weren't cured," Gatenby says.

Around the same time as the breast cancer trials, the father of a colleague of Gatenby's came to the hospital to receive an initial, aggressive round of chemotherapy for lung cancer. According to the colleague, her father arrived on a Friday with no apparent symptoms and was dead by Monday. "That event to me was very traumatizing," Gatenby recalls, and the cause to him seemed obvious. "I couldn't understand why you would treat someone with a fatal disease and kill them with your therapy. It just didn't feel right to me." During this fraught period, Gatenby's own father died from esophageal cancer.

Gatenby felt there must be a better way to treat cancer—to outsmart it rather than carpet-bomb it. He had studied physics in college and believed that biologists could leverage equations to capture the forces driving cancer the same way physicists use math to describe phenomena like gravity.

By 1989, Gatenby was preoccupied with modeling the evolution of cancers. During the day he would scrutinize the X-rays of cancer patients, and at night, after he and his wife had put their young kids to bed, he would sit at the kitchen table in their suburban Philadelphia home and pore over medical journals. The patterns he started seeing in the literature made him wonder: What if cancer cells outcompete normal, healthy cells in the body in the same way an animal species edges out its competitors in nature?

Gatenby recalled that ecologists had come up with equations to describe the balance between predators and prey. As an undergraduate at Princeton University, he had learned the classic example of the math that plotted how growing populations of snowshoe hares fuel the rise of the lynx that feed on them. He began dusting off old books and buying new ones to educate himself on species interactions.

For a year, Gatenby read and mulled. Then, in 1990, on a family trip to the Atlantic coast of Georgia, he found himself stuck in a hotel room one afternoon with his two napping children. Out of nowhere, an idea presented itself. He grabbed a pad of hotel stationery and a pen and began scribbling down some key formulas from population ecology. Those formulas, called Lotka–Volterra equations, have been used since the 1920s to model predator-prey interactions and, later, competition dynamics between species, and were among the ones he had recently brushed up on at home. Gatenby thought this set of formulas could also describe how tumor cells compete with healthy cells for energy resources such as the glucose that fuels them.

When he returned home to Philadelphia, he spent what time he could at a typewriter composing a paper that laid out this theoretical model. As soon as he finished, he showed it to some colleagues. He didn't get the response he had hoped for. They thought it was ridiculous to try to use ecological equations to model cancer. "To say that they hated it would not do justice to how negative they were about it," he says. His peers thought that a brief set of formulas couldn't capture cancer's seemingly infinite complexities.

Louis Weiner, an accomplished oncologist who worked alongside Gatenby at the time, recalls that their colleagues viewed Gatenby's ideas as offbeat. "Treatment orthodoxy at that time favored high-intensity, dose-dense treatments aiming to eradicate every last tumor cell in a cancer patient," says Weiner. "Bob's perspective was antithetical to those beliefs." But Gatenby pressed on and succeeded in getting the paper, chock-full of Lotka–Volterra equations, accepted in the prominent journal *Cancer Research* in 1991.

Despite the publication of his theory, he still couldn't convince oncologists that his idea had practical merit. "I think that they felt intimidated," Gatenby says. "Most physicians are mathematically illiterate." He found that the medical establishment was reluctant to publish much of his follow-up work.

Gatenby held on to his overlooked idea as he moved up the ladder at work. He went on to lead the department of diagnostic imaging at Fox Chase Cancer Center and was later appointed head of the department of radiology at the University of Arizona College of Medicine in Tucson, where he continued to garner recognition for his skilled interpretation of scans. Then, in 2007—the same year that Robert Butler received his initial prostate cancer diagnosis—the Moffitt Cancer Center in Florida asked Gatenby to be the chair of their radiology department. Gatenby had a condition: He would come if the hospital created a division where he could pursue in earnest the link between Darwin's principles and cancer. The Integrated Mathematical Oncology Department, born from this negotiation, became the first math department in a cancer hospital, he says. Finally, Gatenby had a place where he could put his ideas to the test.

. . .

It is, perhaps, surprising to hear that Charles Darwin's theory of evolution is guiding modern cancer treatments. But one of the basic tenets of this approach—that cancer cells can act independently from one another—is apparent in Darwin's own writings from more than 150 years ago. After he published his most famous work, *On the Origin of Species*, Darwin penned a new text in which he pondered what tumors could teach us about the growth dynamics within the body. In that subsequent book, *The Variation of Animals and Plants Under Domestication*, he contemplated the possibility of independent elements within unusual ovarian tumors. The tumors he mentions are almost certainly what are known as teratomas—growths that can contain a bizarre mixture of fully formed tissues, like teeth, often found in other parts of the body. Because of their strangeness, such growths have long captured the interest of biologists.

In later editions, Darwin described odd growths extracted by Lawson Tait, a gynecologist who as a student had become enthusiastic about Darwin's ideas:

> Mr. Lawson Tait refers to a tumour in which "over 300 teeth were found, resembling in many respects milk-teeth;" and to another tumour, "full of hair which had grown and been shed from one little spot of skin not bigger than the tip of my little finger. The amount of hair in the sac, had it grown from a similarly sized area of the scalp, would have taken almost a lifetime to grow and be shed."

Darwin mused that such findings suggested that the smallest units of the body had the ability to operate separately. "Many facts support this view of the independent life of each minute element of the body," he wrote.

What was driving that independence at the smallest scale? What made it possible for cancer cells to go rogue? As we now know, much of what enables tumor cells to become deadly anomalies in the body are the genetic changes that happen within them. It's precisely these genetic changes that Gatenby seeks to rein in and guide with his cancer treatment approach.

The essential insight—that genetic changes underpin cancer—came about well after Darwin had taken his last breath. One big assist came when scientists figured out a cell-staining technique that made cells' tiny bundles of genetic material, known as chromosomes, visible using a microscope. Using this technique, the German pathologist in training David Paul von Hansemann noticed something strange: Most dividing cells would pull these bundles neatly toward two opposing poles as they replicated, but tumor cells sometimes had three or four poles tugging the material in uneven ways. In 1890, Hansemann published his theory that this chromosomal instability underpinned cancer.

Hansemann's idea failed to catch on, but another German biologist, Theodor Boveri, would put forth a similar theory for cancer in a brief 1914 monograph plainly titled *Concerning the Origin of Malignant Tumors*. The time was ripe for Boveri's publication: The concept of the gene and its importance in cellular functions was surfacing. Perhaps for this reason the idea that disorderly migration of chromosomes could precipitate cancer is often attributed to him.

Boveri didn't use the word *mutation* to describe what was going awry in the cancer cells, but it wouldn't be too long before someone else did.

Across the ocean, in the United States, the connection between cancer and genetic anomalies was also starting to solidify. Ernest Tyzzer, who earned his medical degree at Harvard, wrote in 1916 that "it appears logical to regard a tumor as a manifestation of somatic mutation." The word *somatic*, derived from the Greek *soma*, "body," had become shorthand for nonreproductive cells, and *somatic mutations* referred to the genetic changes in all human cells except those relating to sperm and eggs. Tyzzer's thinking was echoed three years later when the scientists Thomas Hunt Morgan and Calvin Bridges in New York wrote that "it is conceivable at least that mammalian cancer may be due to recurrent somatic mutation of some gene." The idea that tumors could be due to faulty genes was born, giving rise to the somatic mutation theory of cancer.

In the years that followed, some researchers began to suspect that aberrations in the genetic code were associated with cancer, but despite heaps of circumstantial evidence, no one had offered direct, consistent proof. Some even thought that genetic mutations were simply the result—rather than the cause—of malignancies. That misconception was challenged by a young doctor named Peter Nowell and his research collaborator David Hungerford.

Nowell had spent the mid-1950s in the navy looking for signs of cancer in the blood of people exposed to radiation in World War II. He had studied at the University of Pennsylvania, and when he returned there four years later, he became part of an effort that leveraged improved ways of visualizing chromosomes within cells. Twenty-three pairs of these bundles of genetic material normally reside inside the nucleus of each nonreproductive human cell. Hungerford, a PhD student who was working toward his doctorate in zoology, was helping investigate the behavior of these chromosomes.

In 1959, Hungerford peered into a microscope and noticed that within the cells of two patients with leukemia, one chromosome—chromosome 22—looked abnormally short. Along with a third scientist, they then confirmed that this genetic abnormality was seen in seven other patients with the same disease. Finally, the world had the evidence it had sought to connect mutation to cancer. The discovery firmly linked a DNA defect with leukemia, and it received wide attention. (Later, it would help pave the way for a revolutionary medication called Gleevec, which has put many people suffering from blood cancer into remission.)

Nowell would go on to put forth a sweeping theory of cancer that had even more profound repercussions. As research continued to show genetic mutations in all sorts of cancers, he saw a crucial commonality. He noticed that among leukemia patients, the sicker a person was, the more chromosomal mutations they seemed to carry. Studies of other kinds of cancers began to show this pattern, too. He started mulling over another theory. If mutations were the source of variation that enabled evolution to occur, then perhaps the mutations happening in cancers were driving a "survival of the fittest" in each patient.

Nowell outlined his theory—that cells inside a tumor are in competition not only with nearby healthy cells but also with one another—in a seminal 1976 paper published in the research journal *Science*. He suggested—and later research confirmed—that certain DNA alterations grant cancer cells resistance against chemotherapy or other treatments, causing them to edge out drug-sensitive cells through a process of natural selection. In the Darwinian battle within tumors, nastier and more malignant cell populations win out over time. Nowell emphasized that tumors may become deadlier as they accumulate more genetic errors. It was an idea ahead of its time. Most scientists back then viewed cancers as the fruit of just a few genetic errors. But Nowell envisioned a universe of mutations within tumors.

Nowell conveyed his radical ideas to his students at the University of Pennsylvania School of Medicine, sometimes smoking a cigarette as he lectured. One of the medical students listening to Nowell lecture in the late 1970s happened to be Bob Gatenby. But Gatenby was years away from arriving at the idea to leverage evolution to fight cancer. That eureka moment would only come more than a decade later.

In the meantime, the scientific community gradually came around to the idea that cancer might contain a complex array of genetic aberrations. It's impossible to ignore the contribution of Bert Vogelstein to this awakening. In the late 1980s, Vogelstein, who had trained as a pediatrician, was leading a lab studying colon cancer. In particular, he and his team wanted to know how tumors grew more aggressive and deadly after they arose.

Colon cancer was an ideal subject for this kind of inquiry. It often begins as a benign polyp. Something happens over time that causes it to become malignant and invasive. Vogelstein and his team dug into what that *something* was. They looked closely and found that mutations in an important gene family called *Ras* were more present in larger growths. Around 9 percent of the benign growths smaller than one centimeter in size had this kind of mutation, but 58 percent of growths larger than one centimeter did. The group went on to find other mutations associated with colon growths becoming malignant. This ultimately gave rise to what became known as the Vogelgram, a model that at the time suggested that a series of four or five mutations could cause healthy colon cells to become malignant. (Later numbers from Vogelstein suggested that the number could be two for cancers affecting the blood, such as leukemia.) There was some later criticism that it was hard to reproduce this theory in other cancers with more complicated pathways, but the idea that cancers become malignant by accumulating mutations stuck.

Scientists had accepted that cancer can be born from genetic mutations, and it was clear that the continuing genetic mayhem inside these growths could shape their ferociousness. In the ensuing years, additional hints emerged that tumors could evolve to become more malignant, as Nowell had suggested. Finally, decades later, Nowell's former student, Gatenby, had the gumption to test whether doctors could manipulate the evolutionary forces within tumors to help patients survive.

■ ■ ■

Sitting in his corner office at the Moffitt Cancer Center in Florida, Gatenby told me that when he was a student, Nowell's ideas didn't make all that strong an impression on him. Gatenby is in his mid-seventies now. His children—the ones who were napping in that hotel room when he jotted down his Darwinian inspiration—now have children of their own, and he has the "I ♥ Grandpa" coffee mug to prove it. Outside his office, roughly thirty scientists and PhD students spend their days researching patterns of cancer growth using equations like those describing population dynamics.

Getting to this point involved a bit of inspiration from the plant world. Around the time that he had arrived at the Moffitt Cancer Center in 2008, Bob Gatenby had come across a historical horticultural discovery that roused his interest. In October 1854, a government entomologist was inspecting some farmland in northern Illinois when he came upon a disturbing scene in a cabbage patch. The large outer leaves of the vegetables were "literally riddled with holes, more than half their substance being eaten away." With each step he took around the ravaged cabbages, tiny swarms of little ash-gray moths rose from the ground and flitted away. This was, it appears, the first record in the United States of the diamondback moth, an invasive pest that in its larval form shows a fondness for cruciferous vegetables. By the late 1800s, the moths were chewing through the leaves of not just cabbages but also brussels sprouts, collards, and kale from Florida to Colorado.

To fight this invasion, farmers started bombarding their fields with primitive pesticides. It worked—or seemed to. Nearly all the moths died, but those that survived the poison reproduced, and the populations of their offspring bounced back stronger than ever. For decades, one pesticide after another failed as the moths evolved to withstand it. Even the grievously toxic DDT was no match for the diamondback. Beginning in the late 1950s, agriculture experts started to abandon the idea of eradication and adopted a new strategy. Farmers would leave the moths alone until their numbers exceeded a certain threshold, and only then would they deploy pesticides. Remarkably, exercising restraint helped. The moths did not die out, but the pest could be managed and the crop damage could be held in check.

When Gatenby heard this history of the diamondback moth, he immediately latched on to it. Gatenby doesn't have a green thumb, nor is he a fan of cruciferous vegetables—in fact, he deeply loathes brussels sprouts. But the story of the diamondback moth appealed to Gatenby as a useful metaphor for his own project to quell cancers. Like the diamondback moth, cancer cells develop resistance to the powerful chemicals deployed to destroy them. Even if cancer therapies kill most of the cells they target, a small subset can survive, largely thanks to genetic changes that render them resistant. In advanced-stage cancer, it's generally a matter of when, not if, the pugnacious surviving cells will become an unstoppable force. Gatenby thought this deadly outcome might be prevented. His idea was to expose a tumor to medication intermittently, rather than in a constant assault, thereby reducing the pressure on its cells to evolve resistance.

Just as ecologists allow for a manageable population of diamondback moths to exist, Gatenby's method would permit cancer to remain in the body as long as it didn't spread further. His name for the approach was adaptive therapy.

At long last, in 2014, Gatenby got permission to test his big idea. He would run a trial of adaptive therapy on advanced-stage prostate cancer patients at Moffitt. The patients had cancer that no longer responded to treatment; drug-resistant tumor cells were winning an evolutionary battle within their bodies, surviving an onslaught of toxic drugs where other cancerous cells had succumbed. The hope was that by using a precise drug-dosing scheme guided by evolutionary principles, doctors could slow the rise of the mutations that endow cancer cells with the fitness to survive.

To Gatenby's knowledge, no one had endeavored to exploit evolution against cancer in a clinical trial until he developed his prostate cancer experiment. He picked prostate cancer to test this approach partly because, unlike other cancers, a routine blood draw for a molecule called prostate-specific antigen (PSA) can offer an immediate proxy for the cancer's progression.

In designing the clinical trial, Gatenby and his Moffitt collaborators had to account for their idea that tumor cells vie against one another for resources. They turned to game theory to plot this dynamic and plugged the resulting numbers for parameters such as growth rates into modified Lotka–Volterra equations. The computer simulations they ran with these equations estimated how quickly drug-resistant cells would outcompete other tumor cells when exposed to the continuous dosage of Zytiga typically given to advanced-stage prostate cancer patients.

In the simulations, the typical administration of the drug led to drug-resistant cancer cells rapidly running rampant. The treatment would ultimately fail each time. That bleak outcome matched up with the results seen in hospital records. In contrast, the computer simulations suggested that if Zytiga were administered only when the tumor seemed to be growing, then the drug-resistant cells would take much longer to gain sufficient advantage to overrun the cancer. Moreover, the equations and digital modeling, along with animal experiments, indicated that when patients' PSA levels fell to under half of the baseline, they could go without Zytiga.

Robert Butler was among the small group of men with advanced prostate cancer who took part in the first small study to test this adaptive therapy. Butler's oncologist explained to him how it would work. He would remain on the Lupron he'd taken for years, and each month he would go to the hospital to get his PSA level tested to judge whether his prostate tumor was growing. Every three months, he would get a CT scan and a full-body bone scan to watch for disease spread. Whenever his PSA level edged above where it stood when he entered the trial, he would start taking the more powerful Zytiga. But when his PSA level dropped to less than half of his baseline number, he could go without Zytiga. These drug reprieves were appealing to him because Zytiga and drugs like it can cause side effects like hot flashes, muscle pain, and hypertension.

The Moffitt approach also promised to be far cheaper than taking Zytiga continuously. When purchased out of pocket, a one-month supply costs more than $10,000. Butler had health insurance, but even so, his first month's supply each year would set him back $2,700 in co-payments, and he'd pay $400 a month thereafter. Going off the drug whenever his PSA level was low would translate to a huge cost savings.

Butler was participating in a so-called pilot trial, which was less rigorous than a large clinical trial, because it didn't randomly assign patients to receive the experimental or standard treatments. Rather, the study relied on a group of patients treated outside the trial as well as results from a 2013 paper on Zytiga to come up with a benchmark for how patients typically fare when receiving continuous dosing of the drug.

When the early results of their new trial trickled in, the Moffitt scientists were gratified and relieved. Ahead of the trial, "we were, to be honest, terrified," Gatenby says. The benefit of adaptive therapy appeared to be huge. Joel Brown, an evolutionary ecologist and one of Gatenby's collaborators, said the team felt a moral obligation to get the word out: "The effect was so big that it would be unethical not to report it immediately," he says.

They published a report in 2017, far earlier than anticipated, to a generally positive reaction from prostate experts—particularly because it suggested a way that people with cancer might live longer with less medication. "Conceptually it's a beautifully simple approach," says Peter Nelson, a prostate cancer researcher at the Fred Hutchinson Cancer Center in Seattle. Jason Somarelli, a biologist at the Duke Cancer Institute, calls Gatenby a pioneer: "He's turning cancer into a chronic disease."

Butler had gone for long periods off Zytiga—with stretches lasting as long as five months. Initially, he was one of the best responders in the study. By 2021, the results looked good overall: Of the seventeen men ultimately receiving adaptive therapy in the study, one left the trial after his disease spread, but most were living longer than expected without their cancer progressing. Data from the comparison group suggested that men getting continuous dosing of Zytiga go a median time of only 14.3 months before the cancer becomes resistant to Zytiga and spreads. In contrast, the median time to progression for the men receiving adaptive therapy was at least 33.5 months, and they were on average using about half the standard amount of Zytiga.

Adaptive therapy doesn't require government approval. The protocol uses already-approved medications, and the US Food and Drug Administration doesn't police specific dosing schedules. Some doctors are already trying adaptive therapy on patients outside of clinical trials. There are, however, caveats to Gatenby's first adaptive therapy trial. The prostate cancer study was very small, and without a randomly assigned control group the results aren't fully reliable. While seven of the men in the trial were still surviving in 2022—including two who remained alive as of 2025—the others have succumbed to their cancer. (Notably, a group of men identified in medical records at the hospital for comparison who were receiving standard care for their prostate cancer all died, suggesting that adaptive therapy had provided a survival advantage to those who received it.) More trials—ones in which patients are *randomly* assigned to be treated with this approach while others remain in a control group receiving standard therapy—are necessary.

Gatenby himself has always been eager for adaptive therapy to undergo more rigorous testing. He conveys a kind of humility you don't see very often in the upper reaches of medical science. He insists that he is not an interesting subject to write about, and more than once I heard close colleagues mangle the pronunciation of his name (which is pronounced "GATE-en-bee"); apparently, he had never corrected them. But when he believes in something, he doesn't relent. And he believes in adaptive therapy. "He's like a teddy bear, but underneath that soft exterior he's made of steel," says Athena Aktipis, who researches theoretical and cancer biology at Arizona State University and has collaborated with Gatenby.

In the years since the start of the prostate cancer study, Gatenby has been trying hard to convince his peers to think differently about the disease. In 2018, he presented his work at a meeting of prostate cancer specialists. In the question-and-answer session afterward, an attendee shared his surprise at the results. "I guess what you're saying is that we've been doing it wrong all these years," the man mused, according to Gatenby. "I was literally speechless for a few moments," Gatenby told me, "and then I said, 'Well, yeah, I guess that's what I'm saying.'" He dwelled on the exchange in the weeks afterward and wished he could somehow find the man and apologize. He didn't want to take back what he had said; he thinks the profession can do better. But, he said, "I should have been more diplomatic."

■ ■ ■

New approaches to fighting cancer must accommodate the frank truth that the disease can be fueled by seemingly endless mutations. Since the early 2010s, improved DNA sequencing tools have shown that Peter Nowell was prescient: Individual tumors often bristle with rapid-fire genetic changes. A significant experiment published in 2012 found at least 128 different DNA mutations in various tumor samples from one kidney cancer patient, for instance. Two years later, an endeavor known as the TracerX project was set up in the UK. It applied new techniques that enabled the DNA sequencing of single cells to reveal the evolutionary history of tumor cells in hundreds of lung cancer patients from diagnosis through treatment. TracerX research has found, for example, that one evolutionary trick cancers adopt is that they double most of the genome inside their cells. Doubling gives them more functional copies of each gene, enabling them to both "experiment" with new mutations and resist problematic mutations that crop up as they multiply with abandon. This variability is one of the harsh outcomes of the genetic instability of cancer: Standard treatments such as chemotherapy and radiation aim to kill cells but often spur resistance mutations that enable those malignant cells to survive.

Robert Gatenby was acutely aware of this paradox. It fueled his desire to see whether he could tip the evolutionary dynamics within cancer toward healthy cells. To prove a key tenet of adaptive therapy, Gatenby and his collaborators in Tampa launched a series of cellular experiments in a lab down the hall from his office. The goal was to prove a key tenet of adaptive therapy. Gatenby's approach assumes that when treatment is removed, drug-resistant cancer cells will replicate more slowly than drug-sensitive cells. The theory rests on the assumption that those resistant cells need lots of energy to maintain their armor against the medication meant to kill them. During treatment breaks, the thinking goes, the fuel-hungry resistant cells are outcompeted by drug-sensitive cells, which need fewer resources to thrive.

To gather evidence for this idea, Gatenby's research team placed human breast cancer cells with resistance to the drug doxorubicin in a petri dish alongside an equal-size population of doxorubicin-sensitive breast cancer cells and watched the two groups fight for resources. By day ten, the resistant cells made up only 20 percent of the cells in the dish and continued to slowly decline from there. At the end of the experiment, these resistant cells had dropped to around 10 percent of the total population.

Granted, this experiment happened in a petri dish, not a human body—or even the body of a lab rat. Some leading cancer specialists agree with Gatenby that drug-resistant cells are likely outcompeted by other cells when cancer medication is withdrawn. But, say others, what if Gatenby is wrong? What if resistant cells actually thrive during the period when the patient is taken off drugs? The risks are high. No one wants to hasten death.

Another emerging matter of concern is that evolution doesn't operate the same way within all cancers. It's not even a given that Darwinian natural selection always determines the genetic mutations that abound within a tumor. A 2015 study of colon cancer samples conducted by Andrea Sottoriva, then at the Institute of Cancer Research in London, and Christina Curtis, a computational biologist at Stanford University, suggested a different pattern.

Bert Vogelstein and his team had suggested the multiple-hit theory in the Vogelgram a quarter century before. But Curtis and Sottoriva were able to use modern genetic tools to see a more complex picture: When colorectal tumors begin to form, there seems to sometimes be a "big bang" of certain kinds of mutations. This initial explosion of cellular diversity in these colon cancers appears to be followed by a period in which random genetic changes arise and become more prevalent out of pure happenstance rather than because the mutations confer some sort of competitive advantage. It's still unclear whether Gatenby's adaptive therapy approach, which operates on the assumption that there's Darwinian competition between tumor cells, would work well for cancers where the mutations arise continuously by chance.

■ ■ ■

We often use military metaphors when we talk about cancer. We battle and we fight, and if we survive, we're victorious. The attitude traces back in part to 1969, when the Citizens Committee for the Conquest of Cancer ran ads in *The Washington Post* and *The New York Times* imploring the president with the words "Mr. Nixon: You can cure cancer." The call to action helped trigger the country's "war on cancer" with a determination that, using enough medical weaponry, the malignant foe could be obliterated. In the following decades, however, it became clear that certain strategies aimed at total eradication were liable to backfire. Blunt and brutal chemotherapy and radiation were toxic, and often lethal.

Rethinking cancer as a *chronic* illness requires a mental shift—a shift that other changes in cancer therapy might be easing. There's a practice of letting cancer patients take doctor-supervised "drug holidays" from their medications, for instance. And we've adapted our thinking before when it comes to medicine. Doctors once thought that stress was the primary culprit behind ulcers, but biologists uncovered a bacterium as the main cause. More recently, we've gotten used to the weird idea that trillions of bacteria live in our gut microbiome.

Perhaps, then, it isn't a huge stretch to think that we might—for certain tumor types—tolerate coexisting with cancer cells as long as we can prevent them from growing unchecked. In that sense, Gatenby is challenging the words emblazoned on the outside wall of the Moffitt Cancer Center: "To contribute to the prevention and cure of cancer." Robert Butler, the former engineer who participated in the prostate cancer trial there, had pondered these words too, when he would walk into the building for checkups and treatments. "Certainly, in my case there's no intention of cure. What we're doing is control. So that's not really the correct logo anymore, is it?" he said to me when I visited Florida in 2018. Butler told me about a time when he and some of the Moffitt researchers had brainstormed alternative slogans. "We finally came up with 'Our aim is to make you die of something else'—which I thought was lovely," he said. "It's more true."

On average, men with prostate cancer in the pilot trial of adaptive therapy have survived far longer than other patients with the disease who have received standard therapy. Now, more than a decade after the start of the study, two men have surpassed the nine-year mark since going on Zytiga and cycling off the drug when their levels of the cancer biomarker protein PSA drop, as prescribed by the evolutionary models that Gatenby and his team devised. Sadly, Butler was not one of them. Around the start of the Covid pandemic, his levels of PSA shot through the roof, hinting that the cancer was spreading further in his body. Butler had become a very beloved patient at Moffitt; he even coauthored a science paper with Gatenby and the other doctors there. But his PSA rose so high that he was disqualified from the adaptive therapy trial, which had preset thresholds. He went on to receive chemotherapy and then entered another trial for a new experimental cancer drug. The side effects of that novel medication were severe, however, and he had seizures following a fall in 2022. He was rushed to a local hospital in June and passed away shortly after.

Gatenby feels the gravity of the loss and is already considering tweaks to his adaptive therapy approach for future trials. One adjustment that he thinks could be effective is to change the drug-dosing protocol. In the original prostate cancer trial, patients such as Robert Butler would go off Zytiga completely for long stretches. Rather than take patients completely off their drugs when their cancer has paused its growth, Gatenby says another approach would be to simply *lower* their dosages during those stretches (and do more frequent check-ins and adjustments afterward). This nuanced approach is particularly valuable for certain types of cancers with a propensity to be forceful. "If you have a very aggressive cancer, you run the risk of losing control of the cancer during the off cycle," Gatenby explains. The new models he and others have created suggest that patients will in this scenario require less and less medication over time to keep their cancer from growing. Other scientists have tested this more continuous version of adaptive therapy in animal studies and found it works well.

Gatenby now also thinks his team might have set their sights too low. His equations had been too pessimistic about adaptive therapy; it might theoretically be able to do more than simply keep cancer in check. The original prostate cancer study in which men cycled on and off the Zytiga assumed that the drug-sensitive cells could never completely outcompete the cells that evolved drug resistance. By this thinking, each time the drugs would be withdrawn and added back, the balance would increasingly tip in favor of the resistant cells. "The math model predicted that we could get no more than twenty cycles, and that we would always lose control eventually," he says.

When Gatenby examined the real-life data from his trial, he saw that with proper timing of drug dosing, the population of drug-resistant cells did not hit a plateau—it actually fell, edged out by the drug-sensitive cells. "If you hit that in that cycle correctly, you could actually get the resistant population to decline to somewhere near extinction levels," he told me. "What the models showed us is that in those guys who are still cycling that's what we did. And in fact, we cannot find any evidence for resistant populations [in those patients]." When his team has looked at the blood of the men in the prostate cancer trial who have survived, they find no evidence of mutated DNA related to the resistant cancer cells. This has caused Gatenby's thinking to change. He is curious about whether adaptive therapy might be able to go beyond keeping cancer at bay to extinguishing it.

In March 2023, an international team of scientists in the Netherlands and Australia announced that they had initiated a new, bigger adaptive therapy trial of 168 men with prostate cancer. Gatenby reached out to the leaders of that prostate cancer trial as soon as he heard about it. He wanted to tell them about his recent updates to the models that suggested adaptive therapy could be tweaked to drive cancer in the body toward extinction. He connected with them and encouraged them to optimize the treatment, taking into consideration all the updates to the approach. Gatenby is constantly refining the adaptive therapy methods. Even as more studies pour in about how treacherous tumor cells' tactics can be, Gatenby has not given up on outsmarting cancer.

■ ■ ■

We go about our days putting on our socks, brushing our teeth, sipping a mug of coffee or tea, and turning the pages of a book, largely unaware that our bodies are humming with trillions of cells. As multicellular organisms, we rely on the cells in our bodies to abide by a social contract and live in harmony. Cancer cells toss that notion aside and replicate with abandon.

"If we look closely enough, every cancer biopsy reveals a history of genetic trials and errors, a record of which mutations helped a lineage of cells survive," says Jeffrey Townsend, a biostatistician at the Yale School of Public Health. "That record lets us quantify the tumor's evolution in real time and—perhaps—anticipate its next move."

With each passing year, we have a better understanding of how malignant cells evolve and persist in patients despite the presence of drugs meant to decimate them. For instance, cancer cells can become impervious to treatments by fusing together and merging their genetic material. In one lab study, exposure to the chemotherapy drug docetaxel caused the number of prostate cancer cells with extra chromosomes to skyrocket from 3 percent to 90 percent. Scientists have suggested that this extra

DNA increases tumor cells' "evolvability"—it essentially gives the cells more tricks up their sleeves—and that the phenomenon is seen across virtually all patients with lethal cancer. Another thing that can go wrong is that certain genes that normally facilitate DNA corrections can become inactivated, allowing malignant cells to amass genetic changes, some of which can give them an advantage. When this happens, "the cancer cells keep generating mutations and each one has the potential to make them less constrained and more dangerous," explains Lynn Caporale, a biochemist and author of the book *Darwin in the Genome.*

We are also learning that mutations enabling drug resistance can emerge in the earliest stages of cancer in the absence of treatment. Tumor cells that happen by chance to evolve drug resistance sometimes lie dormant, emerging to outcompete their drug-sensitive counterparts only when the time is ripe. One study followed patients with medulloblastoma, a kind of cancer that starts in the lower back of the brain. Researchers analyzed tumor samples from the participants before and after treatment. They found that the drug-resistant cells that came to dominate the tumor over time had been present in a tiny part of it from early on.

All this mutation in malignancies is perhaps less surprising when you consider that our bodies may be hotbeds of mutations in cancer-associated genes even when we are healthy. Consider a 2015 analysis of skin cast aside from four people aged fifty-five to seventy-three in a procedure called blepharoplasty—better known as an eyelid lift. None of these individuals who underwent the cosmetic surgery had cancer or any history of the disease. The researchers suspected that the eyelid was a ripe place to hunt for mutations due to UV damage to the skin; it's not a spot that often gets covered by sunscreen. Their hunch was correct. The amount of mutation was staggering: There were more anomalies in a cancer-associated gene called *NOTCH1* found within a five-square-centimeter patch of eyelid skin than seen among more than five thousand cancers in the world's largest database of cancer mutations, the Cancer Genome Atlas. All told, a quarter of the eyelid cells they analyzed had mutations in *NOTCH* genes.

The researchers then turned their focus to the esophagus, the tube that food and drink heads down before reaching the stomach. Once again, they found mutations in *NOTCH1* and cancer-driver genes such as *TP53* in healthy people. In fact, there were more mutations in the cancer-associated genes they examined in the esophagus samples than in the eyelid tissue. In some of the healthy adults, cells with *NOTCH1* mutations had taken over large parts of the esophagus. The scientists were intrigued, and several of them joined a group running mouse experiments to figure out what was going on. In mice, cells with *NOTCH1* mutations were nudging out cells without these anomalies over time. Crucially, mutations in *NOTCH1* also slowed tumor growth in the rodents. These findings support a growing sense among scientists that mutations in *NOTCH1* might help it keep cancer in check, rather than allow the disease to run amok.

Insights like these are causing scientists to shift their views on the importance of studying the evolution of the human genome *within* each individual. "From the point of view of human health, tissue evolution during later life is probably an even more pressing topic than embryonic development," the geneticist Kamila Naxerova of Harvard Medical School wrote in a recent article. "Already, our conceptual understanding of how cancer develops has been shifted profoundly by the recognition that healthy tissues can contain mutations that were previously thought to be relatively specific to cancers." She adds that some of these alterations previously associated with cancer might simply be mutations that tumor cells inherit from normal cells.

The abundance of cancer-associated mutations seems to rise as we age, even if we don't get the disease. Multiple groups have shown that many older people harbor such genetic changes in their blood (more on this in chapter 4). Current methods of analysis have suggested that around one in ten people over the age of sixty-five have acquired mutations in genes associated with leukemia.

Given all these genetic aberrations, why don't cancers occur more frequently? One reason might be that surrounding cells have an influence on the mutational outliers that arise. The notion that cancer can be kept at bay thanks to some suppressive forces within tissues was evident in experiments by embryologists Beatrice Mintz and Karl Illmensee in the 1970s. They discovered that cancer cells introduced into a mouse embryo could be coaxed back to normalcy. Suddenly, it seemed possible that cells that have gone rogue might be subdued by the cells that surround them.

The idea that the body could possibly push back against the forces of cancer intrigued biologist Mina Bissell. She went to work with one of the postdoctoral fellows in her lab to test the theory using a cancer-causing virus. The virus, Rous sarcoma virus, causes deadly cancers in newly hatched chicks, but no cancers arose when they injected it into the *embryos* of chickens. The embryonic cells were somehow keeping the other cells from falling out of line.

Bissell wrote up the results in a paper that appeared in the prominent journal *Nature* in 1984, but her research received a less-than-positive reception from some of her peers. A scientist visiting the Lawrence Berkeley National Laboratory, where she worked, discarded one of her studies when she gave him a copy. "He took the paper and held it over the wastebasket and said, 'What do you want me to do with it?'" she recalled a couple of decades later. "Then he dropped it in."

Despite this pushback, Bissell forged ahead in demonstrating how the area around cells—what scientists call the microenvironment—can influence the behavior of malignant cells. In 1997, she and her teammates showed that human breast cancer cells could be coaxed to become tame when treated with an antibody and put next to normal cells. The cancerous ones became malignant once again when they were removed from that environment.

Over time, experiments such as these have given credence to the concept of a constant "survival of the fittest" among the cells in the body. Understanding those complex evolutionary dynamics, however, is easier said than done. This is all the more challenging when it comes to cancer because tumors appear to have exceptionally high rates of mutations. But researchers have not given up. In fact, scientists such as Bob Gatenby are making headway, with potentially life-extending consequences.

The stark reality is that each of us is a living, breathing consortium of cells. Our multicellularity is a great strength; however, it is also vulnerable to hijacking by greedy players. Among our trillions of cells working together, cancerous ones can emerge and take too much, with devastating consequences. They can mutate beyond control, becoming impervious to our best medicines.

Clinical trials inspired by the adaptive therapy approach aim to trick cancer into turning against itself. At the heart of each is an attempt to tip the evolutionary dynamics within tumors in favor of cells that have not yet mutated to resist drug treatment. It's a race against malignant cells whose nefarious genetic changes equip them to bulldoze past our best available remedies. This new kind of treatment intervenes and influences cellular competition in cancer. But the notion that cell rivalry shapes our health goes beyond cancer. To understand the full reach of this idea, we must begin with the story of a nineteenth-century embryologist who traded swordplay for science.

2.

The Struggle Is Real

"Uniformity is pure delirium."

FRIEDRICH NIETZSCHE

Wilhelm Roux was born to a family that knew how to fight. His father, Friedrich August Wilhelm Ludwig Roux, was a well-known university fencing teacher in Jena, a city within the German Confederation, which had formed after the defeat of Napoleon. Friedrich's own father, Johann Wilhelm, had been a fencing master and authored an instruction manual about the practice based on mathematical and physical principles. And it was more of the same further up the family tree: Johann's father, Heinrich Friedrich Roux, had led a local fencing school in Jena, in his time authoring a manuscript on counterattacking using the right and left hands. It's no surprise, then, that upon Wilhelm Roux's birth on June 9, 1850, his father expected that the newest member of the family would also eventually ascend in the art of swordplay.

Conflict surrounded Wilhelm Roux in his early years. Ongoing territorial battles consumed Jena. Although the city was situated in neutral land during the Austro-Prussian War of 1866, it became subsumed in the Franco-Prussian War four years later. By that time, Wilhelm Roux was a young man, and he briefly volunteered to help the military in medical roles. The war ended in 1871, and his father, perhaps thinking the time was ripe for his son to professionally embrace swordplay, tried to secure a position for him as a fencing master. But fate intervened. The application and book sent with it were mistaken for a library donation and shelved. This accident freed the younger Roux to pursue his own path. Struggle had always been a central aspect of Roux's life, in fencing, in war, and now in thwarting his father's expectations.

Roux described himself as one *"einsamen Menschen"* (lonely human being) and said he had a *"freudearm"* (joyless) upbringing. While the men in his family had wielded swords, he wanted to use a microscope. It's easy to see the rapid transformation in photographs. In one picture from 1870, he is dressed in a service uniform, his eyes staring blankly ahead from under his military cap. Three years later, he is seated comfortably by a table, his hand on a human skull and his eyes now peering through wire-framed glasses.

Roux was living during an era of fast-changing beliefs about human biology. Darwin had published his book *On the Origin of Species* in 1859 and it had been an instant bestseller. This was followed by Darwin's 1868 book *The Variation of Animals and Plants Under Domestication* and his 1871 book *The Descent of Man*. In these texts, Darwin drew on what he had seen during his time on the Galápagos islands. His analysis of finches and the diversity of their beak shapes signaled to him that adaptations to local environments were important in the development of species, and he saw that creatures vied against one another in a process of natural selection for survival. He began using words such as *evolution* to describe the outcome. (The English biologist Herbert Spencer was inspired by Darwin's work to describe the mechanics underlying evolution as a "survival of the fittest," and Darwin, at the urging of fellow naturalist Alfred Russel Wallace, who had independently arrived at the idea of this sort of species change, incorporated this catchy phrase in later editions of his books.) The resulting assertion that humans were animals who had arisen out of this process was nothing short of revolutionary.

It was against this backdrop of big new ideas in biology that Roux pursued his medical training. He studied for several years at the University of Jena—where he had attended lectures by influential naturalists such as Ernst Haeckel, who was publishing theories about how multicellular organisms developed—and was eventually admitted to its medical program. He also spent two semesters in Berlin, studying under a German doctor and biologist named Rudolf Virchow. Virchow believed, as he put it in 1855, that "*omnis cellula e cellula*" ("every cell comes from a cell"). This went against the thinking of the time that cells could come from nonliving matter. (Virchow's work echoed that of Robert Remak, a Jewish scientist who had offered up a similar idea after watching dividing cells under a microscope, but who received less credit.) It seems that Roux was swayed by his mentor, who viewed bodies as a "cellular democracy" or a "republic of cells."

By 1877, Wilhelm Roux had passed his exams to become a doctor, but it was clear he had his sights set on more than just practicing medicine—he wanted to discover how the body formed. Roux was convinced, for example, that he could detect the laws that governed the patterns of blood vessels. To study these patterns more closely, he focused on the human liver. He devised an advanced version of an ancient method of wax injection: Roux would first introduce wax into the blood vessels of livers and then dissolve the surrounding tissue. What remained at the end was a naked casting of the branches. He then turned to math: Roux painstakingly measured the size of the diameter of the blood vessels and the angles at which they branched out from one another. Based on his careful calculations, he concluded that the small branches splitting off the main vessel stem grew out at wider angles than the bigger branches did.

As he surveyed the architecture of blood vessel branching, Roux came to believe that the vascular system developed its structure according to how its components responded to stimuli. The angles at which blood vessels branched off suggested to him that they formed in a way to optimally deliver blood with minimum work. In other words, the functional needs of the body were influencing the way that its anatomy was formed. Roux wrote up his findings in his doctoral thesis, which modern anatomists have called a seminal work in developmental biology. However, Roux was just getting started—he had bigger ideas to share.

Roux was inspired by the work of his teachers such as Haeckel and Virchow, who emphasized that "social" interactions take place among the cells of an organism. And he added a twist. The way Roux saw it, cells of multicellular organisms were not just cooperating but in constant competition. After finishing his dissertation, Roux secured a job as a lecturer at the anatomical institute in Breslau, now known as Wrocław, and began working on an audacious book expanding on his beliefs about the forces that shape human anatomy. The book, published in 1881, laid out a radically different way of thinking about the body, one in which the different elements beneath the skin were engaged in a constant battle. Roux believed, for example, that the tongue pushing against the teeth inside a person's mouth exemplified a competition for space. Mammary glands vying against bones for calcium in pregnant women, he wrote, embodied a competition for resources.

These ideas were neatly summarized in the book's title: *Der Kampf der Theile im Organismus* (*The Struggle of the Parts in the Organism*).

The book leaned heavily on the idea of evolution. This theory had clearly captivated Roux, who referenced his heroes' thinking early on in his own writing. Within the first few pages of *Der Kampf*, he wrote, "Charles Darwin and Alfred Russel Wallace . . . demonstrated that, because of the geometrical multiplication of organisms . . . a struggle must take place among them, with an ever-present possibility of improvement as a result of the constant variation of organisms in all their parts."

But whereas Darwin and his contemporaries had been preoccupied with evolution at the species level, Roux turned the concept of evolution *inward*, beneath the skin. Roux was methodical in how he laid out the Darwinian struggle he envisaged as happening within the body. He described this as occurring on four levels: at the level of molecules, cells, tissues, and organs. (When Roux referred to "molecules," he was not talking about an assembly of atoms, which is what we think of today when we hear that word. Rather, Roux was referring to all sorts of subcellular structures.)

Roux paid special attention to the struggle between cells because cells had the ability to multiply and grow. He stressed, for instance, that no two liver cells are perfectly similar in shape or size, "yet they all fit together in an efficient organ." According to Roux, the differences between cells lay the groundwork for fierce competition. Based on this logic, he ventured that those cells within the body that could "contract strongly and quickly" are selected to form muscles. Likewise, cells with the power to transform matter gave rise to glands, and those that let a stimulus pass through became nerves.

Wilhelm Roux had suggested that cells vie for resources in a way that eliminates the weak ones from tissues. In *Der Kampf*, he wrote:

Because, as we have seen, the individual occurrence as such is not firmly standardized and because from the start not all cells of the same tissue will be of perfectly equal life force, a so-called struggle of cells must occur during the period in which the cells of a tissue are still multiplying; for those cells which, under existing conditions, are best placed to multiply will multiply more rapidly than the others, and thus, with the limited space, more or less take the other's offspring from them, thus inhibiting their further cultivation and proliferation. The more powerful will therefore produce more offspring than the weaker.

Whether it was due to Roux's brief time spent in service during war or his family's fencing obsession, it's clear that he viewed the coexistence of cells in the body through a militaristic lens. As Roux wrote in *Der Kampf,* quoting Heraclitus, "War is the father of all." He viewed the body as a training ground for combat, one in which the best cells rise to power: "When parts struggle against each other to acquire an ever-greater efficiency, the overall performance should also increase," he wrote, "in the same way that the efficiency of an army increases when officers compete and when the best among them are selected to train the novice soldiers." It's clear that he didn't see the competition as a gentlemanly joust but rather as a feature of war.

■ ■ ■

Writing *Der Kampf* was a risk. Roux put forth big ideas, and not everyone was a fan of the book. Its theoretical musings irked one of his former professors, the German anatomist Gustav Albert Schwalbe. Schwalbe offered blunt criticism and added a warning: "Do not ever write a philosophical book of this kind again, otherwise you will never become a professor of anatomy."

But luckily for Roux, his work landed in the hands of some people who appreciated it. The tension between the cells of the body that Roux described piqued the interest of some influential contemporaries. The German philosopher Friedrich Nietzsche obtained a copy of *Der Kampf* soon after it was published and read it closely. Nietzsche used what he learned from Roux's writing as inspiration for his own work, including his doctrine of the will to power, which describes the strong drive for growth and expansion within organisms. According to Thomas Heams, a biologist who writes and lectures on the history and philosophy of science, Nietzsche "is known to have considered Roux's book as his gate toward physiology, a discipline he would develop a special interest for, and to have endorsed Roux's premises, that he once brilliantly summarized in one fragment by the formula 'uniformity is pure delirium.'" *Der Kampf* had captured the imagination of nineteenth-century readers by revealing that the body was abuzz with sundry cells.

Perhaps the person that Roux was most eager to impress with his book was Charles Darwin—one of his heroes. He personally sent a copy to Darwin, who was quite taken with the theories in the text and encouraged others to read it. In April 1881, Darwin wrote a letter to the physiologist George John Romanes, whom he urged to submit a review of the book to the scientific periodical *Nature*. "It is full of reasoning, and this in German is very difficult to me, so that I have only skimmed through each page, here and there reading with a little more care," he confessed. Nonetheless, Darwin held *Der Kampf* in high regard, and told Romanes that, "As far as I can imperfectly judge, it is the most important book on evolution which has appeared for some time."

When Romanes followed Darwin's urging and wrote a review of Roux's book for the journal *Nature* in September 1881, he took issue with Roux for not crediting others who had also theorized about how the survival of the fittest operates within organisms. "Perhaps the most striking feature in the detailed exposition which the author gives of the doctrine is his ignorance of the fact that the doctrine is not original," Romanes wrote. "His work is pervaded by expressions of the importance which he attaches to his idea as that of a new light shining in a dark place, and he is surprised that in the domain of physiology the thoughts of Darwin should not have been applied earlier. But in this country, at all events, the idea is far from being a novel one."

Romanes had a point. As often happens in science, other people were arriving at the same conclusion as Roux. During the mid- to late 1800s, a surprising number of thinkers besides him also touched on the idea that a struggle of subunits could operate within the body. These include—but are not limited to—the Austrian chemist Leopold Pfaundler, the German zoologist Ernst Haeckel, the British philosopher George Henry Lewes, the German English physiologist William Preyer, and the German pathologist Carl Weigert. The seeds of the idea were sown multiple times. But no one had expounded on the notion as explicitly and thoroughly as Roux.

Roux, for his part, seemed at ease staking out the turf of new ideas as his own. Even biographers writing a century later would remark on Roux's self-promotional tendencies. The historian of science Frederick Churchill did not mince his words: "That Roux had a substantial impact upon his peers and upon contemporary embryology there can be little doubt," he wrote, "but it is equally clear that he was a ferocious propagandist for his own accomplishments."

In the years after he published *Der Kampf*, Roux went on to become one of the first people to try to study how the cells of animal embryos divide and develop. In one famous experiment that he published in 1888, he waited until a fertilized frog egg underwent its first cell division, and then used a hot needle to kill one of the two resulting cells. What happened next was unexpected: The embryo would develop a little while longer, but only the right or left half. Roux became involved in an ongoing scientific debate over what determined the fate of embryonic cells, and how they function in concert once they form full-fledged organs.

The ideas about Darwinian forces operating in the body that he had put forth in *Der Kampf* remained popular among many of his fellow biologists for a couple of decades. Alas, that interest would not last. There was an abrupt change at the turn of the century. "After 1900, very few researchers were considering evolution at a sub-organismal level or referred to Roux favorably," the historian Bartlomiej Swiatczak wrote of that era.

If so many people had considered this idea so long ago, then why was it largely forgotten? Swiatczak points to a movement at the start of the twentieth century toward viewing each organism as a harmonious, holistic system, one in which the whole uses the cells. He also notes that another dramatic shift was afoot: The pea plant breeding experiments of the Augustinian abbot Gregor Mendel were rediscovered in 1900. What made Mendel's experiments earth-shattering was that he demonstrated the existence of discrete units of inheritance, which later on would be called "genes." Mendelian genetics put an emphasis on how all somatic cells in the body contain the same genetic material (although experiments decades later would demonstrate this is not entirely true). For much of the twentieth century, scientists only had the tools to see the DNA of cells in broad brushstrokes. They could see the big chromosomes but could not detect the single-letter mutations within them. Partly because of this, the body appeared to be genetically homogeneous. The cells outside the reproductive tract—if they did change—were seen as evolutionary dead ends.

Roux's idea of the struggle of parts within a body faded into history. He never gained the name recognition of his hero Charles Darwin. His book, *Der Kampf,* remained untranslated and neglected in the German that Darwin had found difficult to understand.

In his later years, Roux lived a life of quiet routine. When living in Halle, Germany, he would walk every day at seven a.m. to an island in the middle of the river that runs through the town, regardless of the weather, and have his morning coffee there. He did not like hotels and eventually avoided traveling at all. He would not have known it, sipping his coffee while looking at the rain, but interest in his work would one day be reawakened. Despite being cast aside for decades, *Der Kampf* foreshadowed revolutionary genetic discoveries of our modern era.

■ ■ ■

Almost a century after Roux had published his theories in *Der Kampf,* glimmers of findings suggested he had been onto something. Real, hard evidence for cell competition within organisms came from a very humble creature: the tiny fruit fly. Scientists had previously discovered that some of these fruit flies had mutations that caused the bristles on their backs to be thinner and shorter than their normal counterparts. Because of the diminutive bristles, they were named "Minute" (pronounced "my-NOOT"). An early revelation about cell competition in the fly came in 1975, when two PhD students in Spain, Ginés Morata and Pedro Ripoll, published a paper showing that when cells with the Minute mutation cropped up in the wings of regular fruit flies, those cells would be eliminated. The result was a "surprise," Morata wrote years later when reflecting on the experiment. "We called this phenomenon *cell competition*," he continued, placing emphasis on the concept with italics. It was a manifestation of Roux's theories.

But history repeated itself. After publishing studies on cell competition, Morata, like Roux, ultimately saw interest in the subject wane rather than increase. He and his colleagues had found a striking example of the phenomenon in flies, but few others seemed to want to dig deeper. "Even though there was the general feeling that it was an interesting observation . . . , no progress was made for a long time," according to Morata. One reason why is that scientists instead became captivated by new genetic findings that shed light on the subdivisions of the fly body and how it developed—essentially the genetic basis for the insect's body plan. "Amid all this excitement the phenomenon of cell competition was put aside," he wrote of that time.

In the late 1990s, a couple of new papers emerged "that rescued cell competition from oblivion," according to Morata. They found that isolated cells with mutations in different genes were eliminated within normal fruit flies. The findings inspired Morata to revisit the cell competition phenomenon using updated laboratory techniques. It turns out the mutant cells weren't just disappearing because they were inefficient. They were being actively abolished by other cells through some unknown cellular signals.

It took a couple more decades to finally get more clues about the mysterious cell competition in flies. This is what the scientists found: Cells with the Minute mutation are weeded out thanks to their overproduction of a protein called Xrp1. When they are situated near normal cells, something about their elevated Xrp1 levels sets off a cascade of events that causes them to meet an untimely death. The mechanisms underlying this phenomenon may have implications for cancer. In flies, the gene for Xrp1 seems to have a similar role as that of a tumor suppressor in humans called p53. Some cancer researchers have speculated that a breakdown in the function of p53 might allow malignant cells to skirt the competition that usually keeps mutant cells in check.

It also turns out that, as researchers have said, cell competition comes in different flavors. Cells with mutations can be eliminated, as in the case of the Minute mutation, or they can become "super-fit" and overtake normal cells. Scientists studying fruit flies have discovered more than a dozen mutations that give rise to "loser" cells that are outcompeted by normal cells, and at least as many more that produce "winner" cells that are super-fit. They have also shown that this enhanced competition happens beyond fruit flies. One experiment, for example, offered evidence of super-fit mutant heart cells outcompeting their normal counterparts in mice.

There's also data from rodent studies suggesting that cell competition has the potential to weed out faulty cells at the embryonic stage. Consider what happened when the researcher Tristan Rodriguez of Imperial College London and his team placed different mixtures of mouse embryonic stem cells in miniature lab dishes. All the cells had the capacity to survive, but some were eliminated by competition over the course of four days. Specifically, those that were "unfit"—either because they had defective growth proteins or they had extraneous copies of chromosomes—were edited out as the embryonic cells went through the process of specialization. Other studies of embryonic cells from mice have echoed these findings.

Some people believe that a similar weeding process might also happen in human embryos. They point to unexpected findings from in vitro fertilization, also known as IVF. In many instances, individuals undergoing IVF are given the option of testing the embryos created with their eggs and sperm in the laboratory before those embryos are transferred to the womb. In this "preimplantation" testing process, when the embryo reaches about one hundred cells, a few cells are taken from it and analyzed for their genomic integrity. If, for example, the test detects an abnormal number of chromosomes in those select cells, the embryo is often jettisoned. It's a devastating blow to many couples and individuals who go to great emotional and financial lengths to get a few precious embryos, only to learn that cells from their embryos possess a major genetic anomaly.

Preimplantation genetic testing was first offered in 1990, and it has become a well-established practice. Thousands upon thousands of embryos are discarded each year after they fail to pass muster. However, recent studies have caused some doctors to have a change of heart about whether this testing is necessary when there is no known heritable disease of concern in patients or their families. In 2015, fertility specialists from New York reported that five couples undergoing IVF who had failed to create perfect embryos decided to take the risk of transferring embryos that had been deemed chromosomally abnormal. Two of the couples failed to get pregnant, but the other three gave birth to healthy babies. That same year, a report in *The New England Journal of Medicine* from doctors in Italy detailed the birth of six more healthy babies born under similar circumstances. How could the babies be healthy if the screening had suggested they would have genetic problems? It's thought that the preimplantation tests might pick up on problems seen only in embryonic cells in the outer layer that do not go on to form the fetus. Another possibility is that the embryo might undergo further genetic changes that result in a self-correction. A further notion—one that recalls Roux's ideas—is that normal cells might outcompete the abnormal cells within a patchwork embryo. This cell-cell competition has profound implications for fertility treatment, where genetically imperfect embryos are often cast away.

■ ■ ■

Wilhelm Roux emphasized that no two cells within the same organ were the same and yet they functioned together. Scientific discoveries have borne this out. Cells have niche fates that allow them to take on highly specialized jobs within tissues. This cellular division of labor helps our body function as a whole.

A lot of this specialization happens during development. In the many decades since Roux's frog-embryo-poking experiments, we have learned that as cells mature, they go through a process called differentiation that prepares them for separate jobs in the body. A nerve cell knows to develop a long cable to transmit electrical impulses; a fat cell knows to scavenge for energy-rich molecules and store them inside a large inner sac; and a muscle cell knows to convert energy into contractions, just to name a few. Even within basic categories of cells there is an almost never-ending degree of specialization: Some nerve cells give us a sense of touch and feeling by relaying messages to the brain, while others are specially designed to send signals in the opposite direction, to move our fingers and toes. White fat cells excel at storing calories whereas brown fat cells, which are packed with energy-producing structures known as mitochondria, release heat to keep us warm (and prevent us from shivering). Long, cylindrical muscle cells attach to the skeleton and allow us to coordinate our movement, whereas spindle-shaped smooth muscle cells are involved in the involuntary contractions of our internal organs. Our bodies maintain a careful harmony of cells with highly specific roles.

We can now see how this specificity of roles is reflected at the genetic level: Look inside pancreatic beta cells, and you'll find the genes for insulin production are activated, for example. Meanwhile, in red blood cell precursors the genes for hemoglobin are churning out instructions to make this essential oxygen-carrying protein—and in certain cells of your retina, the genetic instructions for the light-sensitive protein rhodopsin are working overtime. All told, only about a third to two-thirds of the total number of genes in a cell are turned on at any given time. A cell doesn't need to read the entire instruction book of our genome. It simply relies on the genes that will direct it toward its particular role in the body.

Scientists now have the tools to catalog the variety of cells in our tissues with unprecedented detail. Perhaps the biggest undertaking of this kind is a massive multinational endeavor known as the Human Cell Atlas. The seeds of this project were planted more than a decade ago, when computational biologist Aviv Regev and molecular geneticist Joshua Levin were testing new methods of sequencing RNA, the genetic material copied from DNA that carries out the work of cells. The scientists, who worked at the Broad Institute in Cambridge, Massachusetts, realized that they could apply the sequencing technique to detect the unique RNA profiles of individual cells. Regev, Levin, and their other teammates started with eighteen seemingly identical immune cells derived from mouse bone marrow. The analysis revealed starkly different RNA profiles among the cells, indicating that they had different genes switched on or off. From there on out, the project only got bigger. By 2016, Regev had connected with Sarah Teichmann, then-head of cellular genetics at the Wellcome Trust Sanger Institute in Hinxton, UK, and they launched the Human Cell Atlas project, connecting scientists across six continents with the aim of classifying the adult body's estimated thirty-seven trillion cells. The consortium published a paper the next year, proclaiming that "the time is ripe to complete the 150-year-old effort to identify all cell types in the human body."

The Human Cell Atlas grew quickly to include around four thousand researchers from 104 countries mapping cells across eighteen biological networks in the body. They have profiled more than a hundred million cells, including more than sixteen million nervous system cells. Whereas it was previously thought that the body contained only around three hundred different cell "types," studies like the Human Cell Atlas are suggesting that there is significantly more variation than that. In fact, some scientists now prefer to instead talk about cell "states," a term that reflects the fluid identities that cells possess thanks to the different genetic instructions that can be switched on or off within them. By analyzing upwards of nine million lung cells, for example, Human Cell Atlas researchers have identified 144 different cell states that exist at various stages of lung development.

All this diversity underscores that even though we start out as a single cell, we graduate very quickly to a multicellular existence. And what we've learned about differentiation using modern scientific methods echoes what Roux had believed about development. But it's not just the job functions that make cells in the body different. It turns out that if you drill down deeper, you'll find that even cells of the same kind can possess differences at the genetic level.

■ ■ ■

Roux's theories have proven true and important in ways that he never imagined. In addition to preprogrammed differences, another source of variation among cells is random corruption of their genetic code.

Roux, of course, did not know about mutations in genetic code. The structure of DNA would not be discovered until decades after his death. Nonetheless, the emphasis he put on variation within cells of the body foreshadowed what we are witnessing now with modern tools: In recent years, scientists have finally been able to document the dizzying spectra of genetic alterations that endlessly arise in cells of a living organism.

Some of this happens during the constant replenishing of our flesh and bones. The average age of the cells in an adult body is thought to be between seven and ten years. There are cells that hang around for many years—those in the part of the brain known as the visual cortex are typically as old as the person possessing them, suggesting that they do not get replaced. But many others are very short-lived. The cells in the lining of our gut hang around for less than a week, and some in our blood last only a few hours to a few days. By one estimate, a single patch of skin the size of a small stamp will shed around a thousand skin cells every hour.

To keep up with this turnover, the body replaces 330 billion—about 1 percent—of its cells every day, so there are many opportunities for DNA copying errors to happen. As cells divide and multiply, undesirable genetic mistakes sometimes occur in the replicating of their DNA instructions, for example. Those typos in the human genome, which consists of more than three billion letters, accumulate with each progressive cell generation. That's on top of other sources of mutation, such as breakages in DNA strands that are sloppily repaired by enzymes in the nucleus. The mistakes add up over time.

Of the DNA errors that crop up inside us, the vast majority are somatic mutations, which happen from the earliest hours of embryonic development until a person's final breath. Acquired mutations stand in contrast to the genetic errors we inherit from our parents—the typos that were present in either the egg or sperm cell that came together in conception.

The timing of DNA changes influences their reach. Inherited mutations are present at conception, so they are passed on to every cell within a body. They might create variation at the population level, but they are never a source of variation *within* organisms. Somatic mutations are a different story. They are found only in the cells that descend from the solitary one in our body that went amiss. This kind of mutation can happen even in the last decades of life. When it does happen, it can create a subset of genetically distinct cells with an advantage or disadvantage against their neighbors within an organ—setting up an opportunity for a cellular clash.

New studies suggest that the "struggle of parts" Roux tried to convince his peers was taking place in the body affects us from birth to death. The recent findings are thanks in part to novel methods that allow scientists to sequence the DNA of *single* cells within a person's body. Researchers can now differentiate individual cells from one another and understand how they really interact. Single-cell sequencing methods, as well as more sophisticated DNA analysis of bulk tissues, have helped reveal that aggressive cell populations arise all the time—everywhere from the surface of the eyelids to deep within the bone marrow.

There are many diseases in which a Rouxian "struggle between cells" takes place and the bad cells win out. Consider, for example, what happens in Proteus syndrome, a condition named after the ancient Greek god and shape-shifter. In this disease, the rampant multiplication of rogue cells leads some body parts to become oversized, causing severe disfigurement. It was the ailment believed to have afflicted Joseph Merrick, the nineteenth-century Englishman who became known as the Elephant Man. Merrick died at age twenty-seven after suffocating from the weight of his head when he lay down.

Some people have portrayed individuals who are genetic mosaics as medical oddities. But in reality, we all have some degree of genetic differences in our cells, no matter how small and nuanced, piling up over the years within our bodies. "Every human is undoubtedly mosaic," geneticist James Lupski of the Baylor College of Medicine and his coauthors wrote in a 2015 paper on the topic. Given the massive number of cell divisions it takes to form and maintain the body, they write, and how many DNA errors add up in the process, "it is likely that cells exist within all of us harboring countless mutations that could potentially be causative of every human genetic disease."

Thankfully, the emergence of a mutant cell doesn't often doom us. There are numerous reasons why. One scenario is that the mutation occurs in a gene that is not important for that cell's function, so no harm results from its presence. In other situations, its DNA error can trigger a self-destruct program, or the mutant can be removed by the immune system, which constantly surveils for cells gone bad. Yet another possibility is that it might not replicate enough to cause a disease trait to manifest.

At the end of the day, despite our multicellular existence setting us up with constant opportunities for rogue mutant cells to emerge, the upside is that it also dilutes their significance. Trillions of cells make up a human body, and sometimes, it turns out, there is strength in numbers.

The capacity of our cells to mutate from the moment we are conceived—and to sometimes interact in Rouxian competitions throughout our development and lifespan—has consequences in every stage of our growth, in what diseases we battle and in how we age. In my previous writing, I have suggested that we might refer to this inner survival of the fittest as endoevolution—a process of cellular evolution within us.

■ ■ ■

In 2024, more than 140 years after the date of its publication, *Der Kampf* was finally translated into English. It's perhaps no coincidence that this occurred just as the field of genetics was at long last able to show interactions at the single-cell level.

Roux, of course, never lived to see his theories affirmed. But he had long been preoccupied with death. According to one anecdote, he had once promised to donate his brain to science, only to change his mind and rescind the offer. "My brain is small," he said. "Given just a bit of bad luck, my colleagues will believe that you only need a small brain for the mechanics of development," referring to his work in embryology. When Roux died on September 15, 1924, his wife apparently arranged for his body to be injected and preserved with the same methods he himself had perfected. His remains were reported to be embalmed by his successor at the anatomical institute where he had once worked.

Part of Roux's legacy is that the inner survival of the fittest he described also had a positive side. Roux was channeling Darwin's vision of—as evolutionary biologist Stephen Jay Gould put it—"the paradox of higher stability arising through struggle among lower elements." In his book *Der Kampf*, Roux acknowledges this paradox, writing that "to many, the direction of this book may well seem very strange—for it holds that, in an animal, in which everything is so exquisitely ordered . . . a conflict among the individual parts exists" before ultimately concluding that "all good can only arise from struggles." The strife was not just a precursor to the good. It was a prerequisite.

As multicellular beings, we are host to all sorts of microscopic conflicts beneath the skin. Our bodies are buzzing with genetic variation and the constant interaction of cells. Sometimes, as we saw in the last chapter, a struggle can ensue between cells when some of them mutate to become aggressively cancerous. And sometimes those malignant cells win out, with deadly consequences. But Roux's theories remind us that cell competition can also produce good outcomes.

Some of the greatest good to come out of cellular struggles in the body happens within our lifesaving immune system. There is a selection process that transpires among immune cells to anoint the ones that are best suited to battle against each new pathogen. We have such specialized immune cells to choose from thanks in part to the near-infinite genetic variation that is possible within them. Our genetic versatility serves a vital role in ensuring that our immune systems can adapt to new pathogens. As we will find in the next chapter, we owe our life to our defense cells' ability to mutate.

3.

Immunity in Hyperdrive

By the middle of 2021, the Covid crisis had already killed millions of people around the globe and put billions more in lockdown. But there was hope that newly approved coronavirus vaccines would fully protect individuals against the pandemic virus and spell a swift and final end to its spread. In the United States—one of the first countries to have large-scale access to the vaccines—upwards of 70 percent of adults had received at least one shot.

As the months went by, faith that immunization against the virus would help usher a return to normal, pre-pandemic life was increasingly tinged with doubt. People who had received multiple vaccinations against Covid were still getting infected. What good were booster shots, the public wondered, if they could not stop someone from catching the dangerous virus as it mutated? Making matters even more complicated, a new version of the coronavirus known as the Delta variant had unexpectedly taken hold around the world. How well could the vaccines in hand protect against it and future, more deadly variants?

These questions about boosters and variants were front of mind for Michel Nussenzweig when he joined a video call from his office that autumn. Nussenzweig, a physician-scientist then in his late sixties, had spent decades studying how the immune system can evolve new defensive antibodies. As he sat in his office on the leafy sixteen-acre campus of Rockefeller University, perched on the edge of Manhattan next to the East River, he and his collaborators on the video call hashed out an ambitious experiment to find out the true value of booster shots against Covid.

Nussenzweig suspected these shots were strengthening the immune system in underappreciated ways. "People were getting infected with variants, but people who had been vaccinated were not getting really sick with the variant infection," he told me. "We were looking basically for what might be the reason." He had a hunch that the answer would relate to a particular trick of the immune system that allows it to deploy a sort of Darwinian process of natural selection to find the best antibodies.

Nussenzweig collaborated with other lab leaders at the university, including Paul Bieniasz and Theodora Hatziioannou. Together the team recruited forty-two local volunteers who had never been infected by the virus that causes Covid. The participants would donate small samples of blood at the Rockefeller University Hospital at specific intervals after their first Covid vaccine shot and after two subsequent boosters. The scientists would purify plasma from the volunteers' blood, and then expose it to harmless virus-like particles that each mimicked the unique shape of different coronavirus variants, including one that had recently cropped up called Omicron.

The test was designed to indicate the presence of particular antibodies, the Y-shaped proteins produced by the immune system that attach to foreign pathogens in the body and sweep them up for destruction. If the volunteers' plasma could neutralize the proxy particles standing in for the variants, it would suggest they had developed antibodies that could successfully neutralize the corresponding coronavirus variant if they encountered it in real life.

The results offered tremendous hope: More than half of the several dozen Covid-specific antibodies from participants could neutralize the omicron-shaped particles. The finding was remarkable, because none of the vaccines available at the time had yet been updated to specifically inoculate against that variant. Moreover, since none of the participants had caught Covid during the course of the study, the scientists were certain that all the antibodies were prompted by the early vaccines.

Frauke Mücksch, a postdoctoral scientist in a collaborating Rockefeller lab who was heavily involved in carrying out the experiment, felt reassured by the data. "I was just really relieved. I thought, 'Oh, my God—good,'" she says. "I think we all just wanted to share how great the booster is. I definitely told my family 'Get this booster. It's important for sure.'"

The Rockefeller team also analyzed the spectrum of genetic mutations in a few of the volunteers' immune cells. These genetic changes were the driver of their antibody diversity and could give an even fuller picture of their coronavirus-variant-quashing range. Specifically, the team sequenced the volunteers' memory B cells, which act as the immune system's standby squad. These B cells are called on to help generate antibodies when the immune system encounters—or reencounters—a pathogen. The analysis revealed the stunning diversifying potential of these cells. Among this subset consisting of only five people, the tests detected 472 different versions of antibodies against the pandemic coronavirus, a number that underscores the impressive array of antibodies we humans make when faced with a pathogen. The upshot was profound: The detected count of coronavirus-binding antibodies made by those five individuals surpassed the number of prominent variants of the pandemic virus.

The genetic analysis was further proof that each Covid booster shot was broadening the antibody response beyond the version of the virus it was designed to protect against. Those extra immunizations were helping to diversify each person's repertoire of antibodies: Every booster shot would create new populations of memory B cells that possessed a wider range of genetic recipes for the antibodies the body could churn out. "These data help to explain why a third dose of a vaccine that was not specifically designed to protect against variants is effective against variant-induced serious disease," the Rockefeller team wrote in their paper describing their results, which was published in the spring of 2022. As another article published around this time proclaimed, "Boosting begets breadth."

That breadth doesn't happen overnight. Studies during the pandemic confirmed that an immune system response, like a fine wine, takes time to mature. B cells continue to replicate and develop new, varied versions of themselves for many months—and perhaps years—after an offending pathogen is gone (or after an inoculation shot). The findings had implications for the timing of Covid vaccines and boosters. Administering doses too close together in time would perhaps limit the immune response, according to indirect evidence from a study coauthored by Anthony Fauci, who was leading the National Institute of Allergy and Infectious Diseases. Public health officials in the United States began to recommend that boosters be spaced further apart to allow the immune system response to evolve a more diversified arsenal in the months between the shots.

The immune system leverages time to mutate and diversify its antibody-producing cells. It does this to hedge its bets. "The immune system is creating a set of antibodies that not only see the initial thing, but also are sufficiently diverse or sufficiently tuned that they might anticipate something else," Nussenzweig explains. "The immune system is doing an Alice in Wonderland Red Queen thing where it is producing a reservoir of cells that have the potential to see a closely related variant," he adds, referring to a concept in evolutionary biology that species must adapt to stay ahead of continuously changing competitors. The concept is inspired by Lewis Carroll's novel *Through the Looking Glass*, in which Alice is cautioned by the Red Queen that she has arrived in a world where she will have to run twice as fast to move ahead.

Our antibody-producing cells help us stay one step ahead by generating a wide range of mutated subpopulations. This creates a pool of options from which the best ones can be naturally selected to help us when we get infected. Genetic alterations have an important role in equipping us to evolve defenses against invaders such as viruses and bacteria. Mutation throughout our lifetimes is essential for our health. A strong immune system is a diversified one. Unfortunately, not everyone on Earth is lucky enough to have immune cells that can mutate.

■ ■ ■

When Jennifer Allred was pregnant in 2002, doctors gave her the devastating results of a prenatal genetic test. The son she was carrying had inherited an immune disorder from her. In boys, the genetic condition is a full-blown threat. After her baby, who was named Adrian, was born, he started receiving prophylactic antibiotics and antibody infusions to protect him against bacteria and viruses that other infants would easily overcome. His malady also made it difficult for him to benefit from any of the vaccines that children his age are normally given. Vaccines work by prompting the body to cultivate immune memory cells that can store and ramp up an antibody response. But this was essentially a nonstarter for Adrian. The condition he had inherited from his mother severely limited his ability to make such antibodies.

Allred knew the seriousness of Adrian's disease—it ran throughout her family tree. The first sign that something was amiss had been the death of her great-aunt's son in 1944. Baby James had been only twenty-one months old when he passed away. He had suffered from pneumonia, according to his death certificate. Her father, Donald, who was born in 1956, had also been very sick as a child. He would spend several months in the hospital each year with various infections, and during some of the episodes of illness, his parents were told that he would not make it through the night. Around this time, scientists had characterized the disease affecting boys such as Donald, and he was the first in his family to receive a diagnosis: dysgammaglobulinemia. A cousin born in 1967 was ultimately diagnosed with it as well.

The illness is now known by an easier-to-pronounce name: hyper IgM. Patients with this genetic condition can make IgM, which is one of the first antibodies to respond to an infection. But they *cannot* produce other important kinds of antibodies that are more precisely tailored to fight invading microbes. More specifically, their illness prevents beneficial DNA mutations within immune cells that are necessary to generate a broad and adaptable antibody repertoire. Their immune systems remain stuck in place, churning out IgM antibodies but not enough of the others. As a result, hyper IgM patients are highly susceptible to infection.

Allred's dad, Donald, experienced a hodgepodge of infections throughout his life. In his thirties, that included an intestinal parasite called cryptosporidium that wasn't widely known at the time to infect humans (his diagnosis was made by chance when a medical resident, who had brought home Donald's test results, showed them to a veterinarian acquaintance). Ultimately, Donald succumbed to the consequences of a fungal infection caused by aspergillus, a type of mold that is found in soil. Jennifer, who is the oldest of her siblings, was only sixteen at the time.

As more data was collected over time from a growing group of patients, it became clear that there were different forms of hyper IgM. Some forms, like the one affecting Allred's family, result from an inherited error in a protein called CD40L. CD40L normally facilitates a necessary genetic reshuffling in immune cells that helps them switch among the different flavors of antibodies they can make. Other forms of hyper IgM affect the enzyme known as activation-induced cytidine deaminase, or AID for short (which, coincidentally, Michel Nussenzweig has studied). AID is crucial for a process by which immune cells acquire small but vital changes in their genetic sequence to refine the portions of antibodies that help neutralize invaders. This phenomenon is known as somatic hypermutation.

Patients with hyper IgM receive infusions of antibodies to stand in for what their bodies can't make. But there's another option, too: They can undergo bone marrow transplants. The procedure is risky. The initial steps have historically required chemotherapy that wipes out existing bone marrow cells and leaves patients especially vulnerable to infection. But when the transplant works, it repopulates a patient's system with versatile donor cells that can make a complete range of antibodies. Essentially, it can give them a fully functioning immune system. The procedure has become safer and more available as an early intervention, so more families are opting to give it a try, according to Akiva Zablocki, president of the Hyper IgM Foundation. "The story is changing," he says.

In 2003, when Jennifer Allred went with her son Adrian to the hospital in Utah (where they lived) for his bone marrow transplant, the procedure was relatively new for those with hyper IgM. It was a formidable endeavor but also a hopeful one. If it worked, Adrian would be the first member of the family to be cured. Doctors had found a match for Adrian's transplant: a donor from Spain. They started him on intensive chemotherapy to wipe out his bone marrow so that they could later repopulate it with the donor's cells.

The following weeks were particularly grueling. With his immune system wiped out, Adrian had to remain in isolation. Allred stayed with her son in the room, wearing a hospital mask at all times except when she stepped outside the room to eat something. At one point they had to use hair clippers to shave off Adrian's hair because it kept falling out from the chemo.

After the donor cells were infused, there was a lot of waiting to see whether the procedure had worked. When, at long last, the test results came back, they showed that the donor cells had taken hold. After eight weeks in the hospital, Adrian and his mother were sent home. There were plenty of medications to take during the recovery and lots of follow-ups, but the good news kept coming. By day one hundred after the transplant, 92 percent of Adrian's immunity-producing cells were derived from the donor cells. The bone marrow transplant had given Adrian a normal immune system. He now had immune cells that could mutate in a beneficial way to make a full range of antibodies. Adrian is now twenty-three years old and, since shortly after the transplant, he has never needed the antibody infusions that his grandfather, Donald, did.

Insights into proteins involved in hyper IgM are even benefiting people without the disease. Some companies are trying to harness the

AID enzyme to find new therapies for tough-to-treat cancers. The San Diego biotech start-up AnaptysBio engineered cells in the lab to possess AID so that they crank out new shapes of antibodies, just as B cells would do in the body. In partnership with the drug giant GSK, the company used this approach to generate a new antibody therapy that helps destroy endometrial cancer cells. The antibody received approval for sale from the US Food and Drug Administration in 2023.

It's remarkable that AnaptysBio could deploy the AID enzyme in lab cells to come up with a new anticancer antibody. But the miraculous thing is that those of us with healthy immune systems have cells that naturally use such enzymes day in and day out to rejigger their DNA. In this way, among the legion of new antibody shapes they produce, they happen to generate highly specific antibodies that are effective against the vast majority of microbes we encounter during life. The big question, then, is how can the human body so frequently hit the immunity jackpot?

■ ■ ■

Humans, as a species, have many microscopic foes. Our bodies must fend off so many different kinds of viruses, bacteria, fungi, and parasites that scientists have a tough time keeping count. The invaders range from enemies such as influenza and Ebola viruses to *E. coli* and tapeworms. It's not a fun list to think about. All told, there are approximately fifteen hundred different species of microbes on Earth that are *known* human pathogens. But there are likely countless others we have yet to discover. With new and ever-mutating microbes constantly emerging, how does the human body form enough antibody diversity to precisely bind to and neutralize any germs it encounters? And how does it focus its energies on the most immediate microbial danger at hand? These questions have become more pressing in recent decades, with the arrival of previously unidentified pathogens such as HIV, Zika, and the virus that causes Covid, but they were a matter of concern long before these recent microbial invaders came to light.

The discovery of antibodies traces back to the 1800s. The physician-scientist Paul Ehrlich discovered special factors found in the blood that acted against toxins. He dubbed these factors *Antikörper*—which translates from German to English as "antibodies"—in an 1891 paper.

Not too long after, Wilhelm Roux, the nineteenth-century German embryologist, attempted to explain the immune system's versatility. In the second edition of his book *Der Kampf der Theile im Organismus*, published in 1895, he extended his theories of cellular-level Darwinian battles within the body to include immunity. Drawing on the thinking of his contemporary, the pathologist Paul Grawitz, Roux suggested that immunity was acquired thanks to cells in the body that could resist challenges such as infection or poisoning and had a survival advantage. This was a form of "internal rebreeding," as Roux saw it.

Ehrlich, meanwhile, emphasized that resistance to toxins was acquired through a process where the best antibodies capable of responding to these foreign particles were selected by the body for scaled-up production. Some six decades later, others developed theories echoing and expanding on this conclusion. In his 1955 paper titled "The Natural-Selection Theory of Antibody Formation," the Danish immunologist Niels Jerne wrote that the cells that produced the right antibodies at the right time would be favored in the body for a period. This idea of a force akin to natural selection operating on antibodies ignited curiosity in others, including Macfarlane Burnet, a virologist working in Australia. Inspired, Burnet built on Jerne's ideas and formed what he called his theory of "clonal selection." Burnet imagined that when an immune cell produced an antibody capable of successfully glomming on to a foreign substance, that cell would be prompted to start making clones of itself. Both Jerne and Burnet would win Nobel Prizes for their research, but in reality, they were part of a large ecosystem of different labs across the world each putting together a piece of the immunity puzzle. David Talmage, an academic allergist at the University of Chicago, essentially arrived at a theory similar to Burnet's in 1957.

Of course, it's not just a matter of figuring out how the body selects the best antibodies to clone. It also has to have a vast repertoire of antibodies to choose from. In trying to explain the mechanism behind antibody diversity, Burnet turned to genetics. The structure of DNA had been fully deciphered only four years earlier, and there were vast unknowns about how it operated. But Burnet was encouraged by a rising young researcher named Joshua Lederberg to embrace its importance. Burnet eventually saw a possible upside to the mutation that might occur within it. In his 1957 paper, he encouraged other scientists to "picture a 'randomization' of the coding responsible" for antibody components that attach to invading pathogens. By this logic, immune cells' genetic code would get dramatically altered to come up with new antibody shapes that could specifically bind to microbes and mark them for destruction.

Some scientists balked at the idea of mutation as the engine of antibody diversity. They insisted that we possessed a gene in the body for every single shape of antibody we needed to make. But the Japanese molecular biologist Susumu Tonegawa was not persuaded by the latter concept. It simply didn't seem possible to him that the genome had enough room to contain a gene to encode each antibody that could protect against every possible pathogen that existed or could evolve in the future. Using new genetic analysis tools, in the 1970s he and his colleagues found evidence that mutation was indeed the driver of antibody diversity. The experiments revealed that genetic portions of our DNA can recombine in a multitude of ways to encode new antibodies throughout our lifetime. Tonegawa wrote up his findings in the mid-1970s, and—as seems to be a trend for the field—later won a Nobel Prize for his work.

We now know that only a fraction of the entire human genome, which contains around three billion bases, or "letters," of DNA, is devoted to making antibodies. It's downright awe inspiring that this small portion can do so much. Thanks to processes such as recombination, our DNA can render an exponentially high number of possible antibody structure outcomes. Sure, there are approximately fifteen hundred species of known pathogens, but our mutating and remixing immunity genes give us far more antibody variants than that: Our bodies have the theoretical potential to make a *quintillion*—one million trillion—uniquely shaped antibodies. What's even more impressive is that the immune system seems to possess the ability to smarten up about which ones it actually produces.

■ ■ ■

By the time the Covid crisis unfolded around the world, scientists were convinced that we had the ability to make a tremendous variety of antibodies. But they desperately wanted to know more about how the immune system could home in on a new invader. Michel Nussenzweig and his colleagues at Rockefeller University had offered tantalizing clues that the body does not just evolve its defenses in a random direction. This much was clear in the vaccine booster study. With each successive shot, volunteers' immune systems would develop new antibodies to go after parts of the coronavirus outer shell that their existing antibodies couldn't see. The volunteers' antibodies were not diversifying haphazardly—their immunity was smartening up. It must have seemed like a magical process. But it was screaming for a scientific explanation. *How* exactly did the immune system catch what it had missed?

This puzzle gnawed at one of Nussenzweig's lab members, Dennis Schaefer-Babajew. One possible explanation was that the immune system was somehow inhibiting itself from making the kinds of antibodies it already had—and thereby forcing itself to come up with new versions against parts of the virus it had overlooked.

The idea of a feedback loop in the immune system was not new. It had roots going back a century to a pathologist named Theobald Smith. In 1909, Smith showed that giving guinea pigs an excessive amount of serum containing antibodies against diphtheria toxin inhibited the animals' immune response. In the years after, follow-up studies looked for the existence of an internal feedback loop in which antibodies would suppress the production of their own kind. (Unbeknownst to Nussenzweig, his father, Victor Nussenzweig, had co-led an animal study in 1969 that showed that the presence of antibodies against a particular part of a foreign protein accelerated the production of new antibodies that attached to totally different portions of the protein.) Whether this kind of inhibition occurred in humans was anybody's guess. Schaefer-Babajew wanted to look for the phenomenon in pandemic data, but Nussenzweig was reluctant to let his young lab member embark on what he worried would be a fool's errand. "He said, 'I think this stuff has likely been debunked,'" Schaefer-Babajew recalls.

Schaefer-Babajew, a medical school graduate then in his late twenties, thought hard and found a solution to persuade his boss to let him investigate. He remembered that they had stored blood samples from their previous Covid research projects. There was one particular experiment that came to his mind. In addition to their vaccine work, the Rockefeller scientists had run a small clinical trial of two synthetic anticoronavirus antibodies to assess their safety. These lab-made antibodies were designed to neutralize the pandemic virus. They never became commercially available because the virus mutated beyond their reach before they could hit the market. But the blood samples taken in the synthetic antibody trial, which sat deep in the laboratory freezers, were a buried treasure. The participants had received the synthetic antibodies around three months before getting their first dose of Covid vaccine, so scientists could look at sequential samples to see whether this had nudged participants' immune systems away from making redundant natural antibodies of that exact same shape. Schaefer-Babajew persuaded Nussenzweig to green-light the investigation by emphasizing they could simply reanalyze these samples they already had in hand.

Schaefer-Babajew's hunch that present antibodies could nudge the body toward making new, different ones proved accurate: Study participants who had received the synthetic antibodies prior to vaccination possessed a more diverse repertoire of their own antibodies after that shot than those who had not been given the infusions. Additional data from the experiment shed light on part of the reason why: The synthetic antibodies had clung to and hidden specific parts of the coronavirus shell, likely prompting the participants' immune systems to go after other bits of the virus's outer layer.

Our own natural antibodies might participate in the same kind of masking phenomenon when our bodies make high amounts of them. This could be how we gradually evolve better versions of antibodies to attack different parts of invading pathogens even without vaccines.

And when an immune cell makes an effective antibody against a pathogen, it then produces clones of itself. Each of those clones is capable of churning out that winning antibody.

It's a beautiful system, fueled by mutation. We cannot survive without it. But it is also a system that can accidentally evolve in the wrong direction. In some cases, immune cells mutate to attack our own tissues. Among the quintillion theoretical antibodies that the body has the mathematical possibility to evolve, it somehow begins producing ones that go after its organs. And when these rogue cells escape the checks and balances meant to weed them out, the results can be absolutely devastating.

∎ ∎ ∎

When Christian Mayer joined Nussenzweig's lab after finishing his PhD in Germany, he was deeply interested in studying autoimmunity, the misdirected attack by the body on its own cells. As a teen, he had watched the autoimmune disease rheumatoid arthritis overcome his aunt. "At that time, treatment options were very limited," Mayer recalls. "I saw how basically every joint in her body was destroyed. She died in her early sixties from the disease." Mayer wanted to make a difference.

Part of the challenge of understanding autoimmunity is knowing how the constantly shifting immune system can take a wrong turn. The evolution of your immune system is happening all the time. As you read this sentence, there are cells deep in your bone marrow called lymphoid progenitors that are giving rise to various immune cells. While still in the bone marrow, some of these cells begin undergoing a series of developmental steps—including the remixing of their antibody genes—to become B cells. Those B cells are then released into the bloodstream and end up maturing in various places in the body, including the spleen and lymph nodes. It's in sites such as these where their antibody genes are mutated for precision. But there are times when genetic alterations don't advance B cells toward a better antibody. In fact, there are times when they even mutate to turn against the body's own healthy tissues. This is the basis of autoimmune disease.

The idea that the immune system can evolve in a self-sabotaging direction has concerned scientists for decades. Just a couple of years after publishing his 1957 paper on his clonal selection theory of immunity, Macfarlane Burnet put forth his "forbidden clone hypothesis." Burnet made the case that the immune system must weed out "forbidden" populations of cells that attack the body or else risk developing autoimmune disease. That editing turns out to be a mightily complex process. For starters, the immune system churns out a lot of bad apples. Scientists now know that approximately 40 percent of early immature B cells in healthy humans are promiscuous in what they bind to or react directly against the body's tissues. Many of these rogue B cells are culled on their journey of maturation. But when a self-reactive cell slips through and starts multiplying, the resulting autoimmune disease is hard to reverse.

Autoimmunity is a more common affliction than you might think. By some estimates, 5 to 10 percent of the general population in Western countries develop some sort of autoimmune disease. These disorders range from celiac disease to type 1 diabetes to lupus. To make matters even more urgent, doctors have sounded the alarm that autoimmune diseases appear to be on the rise.

Mayer shared his interest in understanding the growing problem of autoimmunity with Nussenzweig. Eventually, Nussenzweig convinced him to look deeper into how and where the body's self-reactive cells were culled. "He told me, 'This is a really exciting question, and you should develop a tool so that we can address it,'" Mayer says.

Mayer got to work and developed a special type of mouse; he engineered the rodents so that certain immune cells would emit fluorescent light depending on whether those cells were alive or dying. This emission could indicate what happened to the cells when they passed through so-called "germinal centers" in the body. These centers are where their mutations are normally refined. It's thought that the self-reactive cells that bypass this checkpoint and don't die or get corrected can continue on to cause autoimmune disease. Depending on whether the cells in the rodent study fluoresced, Mayer could tell if they had survived the germinal centers. He and his colleagues found that some cells making self-reactive antibodies did indeed escape this surveillance. But there was an extra twist: Those rogue cells would sometimes undergo a type of programmed death afterward. This suggested an additional, underappreciated way that autoimmunity is naturally kept in check.

The fact that the immune system has both surveillance and self-destruct mechanisms to weed out bad actors points to the high risks of mutations gone wrong. And the risks might be layered: Epidemiological studies have suggested that people who suffer from autoimmunity might also be at greater risk for developing immune cells that mutate to become cancerous. This hints that some individuals might have immune systems that either mutate too much or fail to delete cells that have gone awry.

Perhaps if the immune system were not so reliant on mutation as the engine of its antibody defense system, it would never accidentally evolve self-destructive B cells. But it must mutate, and so some of the molecules that help it evolve are also the ones implicated in its mistakes. For example, one theory is that the mutation-causing enzyme AID—which gives us a vital antibody repertoire—can sometimes be a culprit in precipitating bad genetic changes. "We think that autoimmunity is the price we pay for the diversity that AID gives us," says Mayer, who is now a researcher in the US National Institutes of Health's Experimental Immunology Branch. Some mouse studies have supported this notion by showing what happens to autoimmunity in the absence of AID. When a strain of mouse prone to developing a lupus-like disease was engineered to lack the enzyme, they had less kidney damage and lived longer.

To paraphrase Glinda of *The Wizard of Oz*, it's difficult to know if AID is a good witch or a bad witch. If AID were a big culprit of autoimmunity, one would expect people with defective or inactive AID to have a lower risk of autoimmune disease. Paradoxically, that doesn't seem to be the case. People with genetic conditions who lack normal AID function—like some with hyper IgM syndrome—are more prone to a gamut of autoimmune illnesses, such as the gut ailment known as Crohn's disease. In light of this data, scientists now suspect that both overabundance *and* underabundance of the AID enzyme might elevate the risk of the body evolving self-attacking antibodies.

Modern medicine has striven hard to counter the immune system's dangerous duds. In the past several decades, the treatment of many autoimmune diseases has been revolutionized by a class of drugs called monoclonal antibodies. These lab-made antibodies can essentially go after rogue cells that make self-reactive antibodies or other harmful chemical signals. They have helped patients but get marketed at sky-high prices that are beyond many people's reach. The most lucrative drug in history is Humira, a monoclonal that treats various conditions, including rheumatoid arthritis. In 2021 alone its global sales reached around $21 billion. The only drugs that beat it in sales that year were Pfizer's and Moderna's Covid vaccines.

For certain autoimmune conditions, doctors will sometimes even use a process called plasma exchange therapy. In this approach, they run a patient's blood through machines that separate out its watery component, called plasma, which contains the self-reactive antibodies. They replace this with other fluids—sometimes donor plasma—so that the blood that flows back into the patient isn't as damaging. This method is used to treat conditions such as Guillain–Barré syndrome, in which some of the body's antibodies start attacking the nervous system.

The ingenuity that has given doctors the ability to intervene when the immune system goes wrong is impressive. Modern medicine now includes drugs that can remove cells that make a bad antibody, and in doing so can improve and even save lives. But what about the flip side? Instead of subtracting a particular destructive antibody from our blood, how about fast-tracking the emergence of a beneficial one? Vaccines can prompt the body to try to go after a given pathogen, and boosters can encourage a diversification of antibodies, but could we somehow steer the immune system to produce the exact protective antibody we wanted? It's not just a pie-in-the-sky dream. A small group of researchers have been part of an ambitious effort to give our cells a shortcut to evolving precisely shaped super-powerful antibodies.

■ ■ ■

For many immunologists, an effective vaccine against HIV is their white whale. It seems forever out of reach, but so important to achieve. "The question that everybody always asks me is, 'You've been working on an HIV vaccine for thirty-five years and scientists made a Covid vaccine in eleven months. What's your problem?'" Barton Haynes says. The answer has a lot to do with the different ways the viruses behave in the body and our ability to keep up with the right antibodies.

Haynes, an immunologist at Duke University, has long been fascinated by how the body might—somehow—come up with the right antibodies against HIV. Two decades ago, he traveled from North Carolina to Johannesburg to help lay the groundwork for a new kind of study that explored this possibility. Many of the previous research trials of HIV/AIDS—not to be confused with the AID enzyme, mentioned above—had followed a couple hundred to perhaps a thousand people at most. But Haynes had received a new grant to embark on an HIV study of a much grander scale. "Our mandate was to set up sites all over Africa, and do it differently than anyone had ever done before," he says. He and his collaborators connected with more than a half dozen clinics across several sites in Africa treating sexually transmitted diseases and recruited study volunteers. The broad reach of the project meant that it included some participants who had only recently become infected. Their bodies hadn't yet had time to mount an immune response against the virus. "We tested thousands and thousands of people when they came into study clinics and found people who were virus-load positive and antibody negative," Haynes explains. The researchers were able to track the initial antibody evolution against HIV happening in these individuals. This was just the beginning, though. Haynes and his colleagues continued testing the volunteers' blood again and again, over a span of many years.

Among the HIV-positive individuals in the trial, there was one man whose body had begun to make a fairly decent antibody against the virus. That piqued the interest of Haynes and his collaborators. When they sampled the man's blood periodically over the course of more than three years, they saw he was gradually making even better ones. Eventually, he made a remarkable antibody called CH103, which could neutralize a broad range of HIV variants.

Around that time, virologists were also publishing stunning data showing that HIV would mutate and evolve extensively inside each individual it infected. Haynes and his team had a lightbulb moment. They realized they could plot the genetic shifts of HIV within their unique study participant against his antibody changes over time. They hoped this side-by-side comparison would reveal which bits of HIV sequence helped nudge his immune system in the right direction. The researchers retrieved the stored blood samples from the man who made CH103 and reanalyzed them to see how the copies of HIV in his system had changed over the study period. "We said, 'Oh! The broadly neutralizing antibody developed in that person,'" recalls Haynes, who has a trim white beard and a slight Southern lilt. "The blueprint must be in those sequences." It was infeasible to deliver antibodies as therapies at a large population level—and they endure for only a limited time. Scientists wanted to figure out how CH103 evolved in the man so that they could develop a way to generate antibodies like it with vaccines, which are much easier to deliver and typically produce lasting immunity.

The scientists saw that an arms race had happened within his body. HIV would outsmart his antibodies, but then his immune cells would produce even better neutralizing antibodies, which the virus then mutated to resist, and so on. This iterative process is what ultimately produced CH103.

The discovery of CH103, published in 2013, came at a critical juncture in HIV research. Major HIV vaccine trials had failed, and the field was grasping for insights that could correct its path. The CH103 finding gave tremendous hope that it was possible for a person to make a potent antibody against HIV. Now the challenge was to help other individuals achieve this as well.

Haynes was game to try. He had been working on AIDS since the early 1980s, when the cause of the disease was not understood. At the time, some people questioned the value of his work. A few even refused to shake his hand out of fear of contracting the illness from someone studying it. Finding CH103 encouraged him that the fight against HIV was not lost. He and others had begun to imagine a radically different kind of vaccine against HIV—one that tried to shape the evolutionary arc of antibody maturation in the person receiving the shot. Traditional vaccines were designed with zero regard as to how a person's immune cells mutated to make a protective antibody. The new approach would target B cells near the beginning of their evolutionary journey and—by exposing them to the right fragments of HIV in a carefully ordered series of shots—nudge them toward so-called "improbable" mutations that generate super-antibodies. This strategy, which aimed to re-create the evolutionary lineage of potent antibodies like CH103, was dubbed "lineage-based vaccine design." Scientists planned to develop this treatment for people living with HIV, but they also wanted to apply it prophylactically to uninfected individuals to equip them with super-antibodies that could block the virus from taking hold in the first place.

Lineage-based vaccine design is especially appropriate for HIV because of how particularly tricky the virus is. Unlike coronaviruses, HIV can adeptly insert itself into a person's DNA and from there constantly spew out new mutated copies of itself. "HIV is an integrating virus and can hide from the immune system, whereas the virus that causes Covid is not integrating like HIV and can be controlled by the immune system," Haynes says. "It's a huge difference. You don't need lineage design for a Covid vaccine." Moreover, HIV's propensity to mutate inside its host is unmatched. The diversity of HIV sequences in a single person infected with the virus who doesn't receive treatment is about the same as the diversity of all the influenza sequences in any given year within the worldwide human population. HIV simply mutates too fast and too much for the average person to evolve a protective antibody in time. When such broadly neutralizing antibodies do arise, they do so only years after infection, in a small minority of patients such as the man who made CH103. Even when this happens, it's typically far too late to change the course of the disease. Lineage-based vaccines promise to create a fast track to those much-needed super-antibodies.

It's true that vaccines for Covid and flu are given periodic updates, but they are tweaked only to mirror the dominant strains of their corresponding viruses. The evolutionary trajectory of a recipient's immune cells is not taken into account. By comparison, lineage-based vaccines consist of a series of painstakingly designed shots that show the right bits of the virus to the immune system in the right sequence to guide the mutational path of the antibody-producing cells.

The lofty idea of lineage-based vaccine design got a big boost from animal studies. These include an experiment co-led by Michel Nussenzweig. In a paper published in 2015, he and his colleagues reported that they successfully coaxed mice to make broadly neutralizing antibodies using a two-step process. A few years later, Haynes published research showing similar success with tailored immunizations in macaques.

Finally, Haynes joined forces with more colleagues to test a lineage-based vaccine approach in humans. They got a small pilot trial off the ground in 2019. It followed twenty-four participants, twenty of whom received a vaccine designed to tip naive B cells toward the kinds of mutations that would create broadly neutralizing antibodies. The remaining four received placebo shots. Partway through the trial, one of the participants had to be treated for an anaphylactic reaction several hours after receiving the treatment. As a result, the study was halted. But even without all the volunteers receiving their complete series of shots, the study offered encouraging clues that the experimental therapy had begun to work. One of the antibodies that a participant produced was able to neutralize an impressive number of different HIV strains. In fact, as they reported in 2024, it was able to work against 35 percent of the very toughest-to-thwart ones. For Haynes, it marked a milestone in his decades-long work on HIV vaccines. "After thirty years this is the first such success in humans," he told me.

One thing that has become clear to Haynes and other immunologists in recent years is that a vaccine will need to induce *multiple* kinds of broadly neutralizing antibodies to confer protection against HIV infection. Making a single type of super-antibody won't cut it. So, while the pilot trial aimed to produce antibodies against the gp41 protein on HIV's outer shell, the trial they launched in 2023 seeks to generate antibodies that tackle another of the virus's proteins, called V3-glycan. At the moment, many groups are exploring lineage-based vaccines against HIV. After nine years of refining one such approach, researchers at the Fred Hutchinson Cancer Center in Seattle began a recent clinical trial. The International AIDS Vaccine Initiative has also sponsored trials, including one in collaboration with the biotech company Moderna (famous for its mRNA Covid shot).

The more scientists have learned about the remarkable potency of broadly neutralizing antibodies, the more they are set on prodding the mutational path of B cells with vaccines so that anyone will be able to produce these stellar antibodies. The urgency with which they want to apply this to HIV—which claims more than half a million lives worldwide each year—serves as a stark reminder that understanding the evolutionary forces in the immune system is more than an academic pursuit. It is knowledge that, if applied correctly, could potentially save millions of lives.

With each passing decade, it has become increasingly evident that genetic changes are a crucial component of a functioning defense against pathogens. In the immune system, mutations, which have famously been called the "raw materials of evolution," are a necessary force for good. Our antibody-producing cells essentially learn over time, and when they have a winning attack against an invader, they produce clones of themselves.

Once you grasp the importance of clones in the body, it's hard to look back. There's something astonishing about a highly specialized immune cell saving us from sickness by essentially making Xerox copies of itself. But it's not just the immune system in which cellular clones have a role. Only a couple of decades after Macfarlane Burnet published his theory about the clonal selection of antibody production, the pathologist Peter Nowell wrote a paper entitled "The Clonal Evolution of Tumor Cell Populations." In both cancer and immunity, scientists had begun to see Darwin's principles in action. This, however, was only the beginning. As the world started waking up to the important ripple effects of cellular clones, a young Italian doctor would make a surprise discovery about them in blood.

4.

Attack of the Clones

"The dream of every cell is to become two cells."

FRANÇOIS JACOB

In 1964, Lucio Luzzatto packed up his belongings and flew with his wife and four-month-old baby from Italy to take a position as a hematologist in a Nigerian university hospital. It wasn't too long after that when, during one of his hospital shifts, he received a worrying phone call. A local Nigerian obstetrician, Adebayo Oni, had rung to express serious concerns about a pregnant patient with anemia. "He said, 'We see so many pregnant women with anemia. But I have a feeling that she's different,'" Luzzatto recalls. The patient, a twenty-six-year-old seamstress, had experienced anemia before delivering each of her first two children, but five months into her current pregnancy she had a particularly odd symptom: During the night, she would pass dark red urine.

She was very weak, so the medical team started her on antimalarial medication as a precaution, along with iron and folic acid supplements. Yet her health did not improve. She vomited and complained of abdominal pain. Luzzatto kept running tests, trying to find the source of her illness. Then he struck upon an important clue: When he tested her urine, he found free-floating hemoglobin, the pigment-loaded protein that shuttles oxygen and gives our blood its red hue. More tellingly, when he incubated her blood cells with a special serum and then spun the test tubes to separate the liquid, it was pink. In a healthy person, the liquid remains clear. The pink color indicated that the patient's blood cells were likely being destroyed by her own system and releasing hemoglobin in the process. With these results in hand, Luzzatto was able to make a diagnosis: The pregnant woman had a condition called paroxysmal nocturnal hemoglobinuria, or PNH for short.

The strange disorder afflicting her had a long history of baffling doctors. All the way back in the late 1600s, a high-status gentleman in Danzig—the largest city on the Baltic seaboard at the time, now known as Gdańsk—complained that sometimes, for no apparent reason, his urine would turn red and then nearly black. Worse, when this happened, he would feel as though he had a tight belt wrapped around his waist. Two centuries later, in 1866, a leatherworker in London complained of red urine and a severe "pain in his loins," according to his doctor's report. He was admitted to the hospital and remained there for nearly two months. Some mornings his urine would have "a red Burgundy-wine color," whereas other times it was completely normal. Ultimately, he was well enough to leave the hospital, but the mystery surrounding his illness—and the other cases like it—lingered.

By the time Luzzatto had arrived in Nigeria, doctors knew this disorder as PNH and understood that it could cause urine to turn dark because of the hemoglobin that spilled out of red blood cells. But uncertainty remained about how the disease would crescendo within a patient. And PNH still seemed exceedingly rare. It was believed that PNH affected only around one in five hundred thousand people. Luzzatto had never seen a case of it in the course of his relatively young medical career.

Despite his and the other doctors' best attempts to treat the pregnant patient, her health continued to decline, threatening not only her but her child. Luzzatto was desperate to understand the root cause of her PNH. "When we diagnosed her, suddenly I became obsessed," he says. He pored over medical journal articles to try to find some clues. One thing that caught his attention was a theory articulated in 1963 by the British hematologist John Dacie. Dacie was world renowned for his expertise in anemia disorders, and he posed a provocative question in an article: "Is it possible that PNH occurs as the result of somatic mutation . . . in the bone marrow?" In other words, was PNH the result of a mutation in bone marrow cells that caused them to make defective copies of blood cells? Dacie also believed that a mutation could give these copies, or "clones," a leg up. He wrote that "the abnormal cells must have some, as yet not understood, biological advantage."

Luzzatto was instantly intrigued. "It occurred to me that the clonal idea was the best one, but it was not proven," he says. With samples from the pregnant patient in hand, he devised a test that no one had tried before.

Luzzatto took advantage of a medical coincidence: In addition to having PNH, the patient also happened to be a carrier for a completely different condition caused by a mutation on the X chromosome. He used that X-linked condition as a proxy to track the populations of her blood cells. Usually, a woman's body will silence one of the two X chromosomes in each cell at random throughout the body, and it is said to usually average out so that each copy is silenced in about half of the cells. But when Luzzatto incubated the pregnant patient's blood cells, it wasn't a fifty-fifty split. This skewing hinted that there was a clonal overgrowth of cells happening in her blood, just as Dacie had predicted.

Meanwhile, the medical team was trying everything they could to save the pregnant seamstress and her unborn child. They gave her a blood transfusion, and then another one, but everything quickly turned for the worse. She developed a fever and went into premature labor. Her baby boy lived only a few minutes. "It was very sad. Very sad," Luzzatto recalls. Four days later, the patient herself passed away. An autopsy revealed that, in addition to PNH, she had a parasitic intestinal infection, which had likely precipitated her death.

Luzzatto did not want others to suffer as she had, so he kept searching for answers. He and his colleagues identified a second woman with PNH who had the same unrelated X mutation as the pregnant patient. Her blood cells were also skewed. Importantly, in the second woman, the cells in which the normal X chromosome was silenced grew faster than those in which the mutated X chromosome was switched off. The opposite had been true of the cells taken from the pregnant patient. This confirmed that the X chromosome mutation was not related to PNH. Something else was helping one population of blood cells win out in each patient. They would later figure out what that "something" was, but in the meantime the experiments gave the much-sought evidence that PNH was caused by cellular clones taking over in the blood.

Luzzatto worked with Oni and another colleague to write up their findings in hopes that it could convince others that PNH was a "clonal" disease. Perhaps this discovery could pave the way to new treatments, they thought. As a young doctor just starting out in his career, Luzzatto was elated when the paper was accepted and appeared in a research journal in 1970. "I never thought they would even publish it, but they jumped at it," he says.

The discovery of PNH's clonal nature opened Luzzatto's eyes to the way competition and evolution operates within the human body at the cellular level. Darwin, he says, had the right intuition that certain biological events—what we now call mutations—make evolution possible. "What Darwin could not anticipate is how clearly [mutation and selection] operates not only in populations of organisms but also in populations of somatic cells," Luzzatto wrote. His confirmation that PNH was a disease of cellular clones illustrated how the mutant cells that seed the landscape of our bodies can flourish at our expense.

In the years following Luzzatto's early experiments, scientists learned that there are a couple of stages in the development of PNH. One key event is a spontaneous mutation in some blood stem cells within the gene for a protein called PIG-A. Cells carrying the mutation for PIG-A somehow grow to outnumber healthy blood cells. This creates a dangerous situation because the mutant cells that take over are also prone to demolition by the immune system. That's because they lack a proper version of PIG-A, which normally protects red blood cells from being destroyed. So a vast number of the blood cells start dying off. As a result, a person with this stage of the illness will have a low red blood cell count and other complications.

After Luzzatto and his peers discovered the role of clones in PNH, it became a prime example of a genetic disease caused by a spontaneous mutation rather than by an inherited one. It suggested it wasn't only cancer and immunity where such mutations mattered. In reality, PNH was just the tip of the iceberg of a huge range of health conditions linked to mutant cells in the blood.

■ ■ ■

When Patricia Jacobs and her colleagues looked at the magnified cells they'd collected for their study, it was clear that something was terribly wrong. This was in the 1960s, long before modern DNA sequencing.

At the time, one of the main methods that scientists had to assess the genetic integrity of a cell was to simply stain it with a dye and then peer into a microscope to count how many chromosomes it possessed. The chromosomes appeared as deeply hued squiggles on the glass microscope slides. Each cell in regular tissue in the body normally contains forty-six chromosomes. Jacobs, who worked at the UK's Medical Research Council, had every reason to expect this number when she evaluated blood cells from almost a hundred healthy people ranging in age from five months to eighty-two years. None of these study participants had ever received X-ray treatment, which can cause genetic damage. Their chromosome count should have been normal. But when she and her collaborators analyzed the cells, they saw a strange trend: Some of them had entire chromosomes missing, and this was especially true of those from older individuals.

Two years later another study came out from Jacobs and other scientists at the same research institute, including the epidemiologist Richard Doll, who famously linked smoking to cancer. In this follow-up, the team determined that the chromosomes people would lose in some of their cells as they got older were often their sex chromosomes: In other words, women would lose an X chromosome and men would lose their Y chromosome. The latter, known as "loss of Y chromosome," received more attention, which is perhaps unsurprising given that medical research has historically focused more on men.

Within a couple of decades, scientists began to connect loss of Y with disease. The first study to do so was published in 1985. In the experiment, researchers examined the genetic integrity of bone marrow cells taken from patients with a kind of cancer called acute myeloid leukemia. They found that loss of Y was present in the patients they examined, and the condition seemed to revert back to normal in those whose cancer was successfully treated.

It's tempting to see data like that and presume that losing the Y chromosome causes cancer. But it's best not to leap to that assumption.

The truth is, chromosomal loss might simply be a red herring. A study of genetic data from more than two hundred thousand men in the UK uncovered a twist: It seems that *other* gene mutations might be at the root of *both* loss of Y and cancer. The analysis found more than 150 DNA variations associated with Y-chromosome loss and discerned that many of them were located near cancer-susceptibility genes. Notably, the variants linked to loss of Y were also correlated to cancers affecting the brain, kidneys, and other organs in both the men *and women* in the study. Women do not have any Y chromosomes to lose, so it's unlikely that the disappearance of this chromosome is to blame for the malignancies. Scientists believe that DNA changes such as the 150 variants they identified might make a cell prone to genetic instability—which, in turn, could result either in the accidental discarding of a chromosome or in DNA errors that turn a cell malignant. In other words, perhaps loss of Y and cancer simply share a root cause.

Despite the uncertainty about whether loss of Y is a cause of disease or is simply along for the ride, interest in the phenomenon has continued. Loss of Y has now been associated with a myriad of problems ranging from diabetes to Alzheimer's disease to a shorter lifespan.

There's also a better understanding of how frequently this chromosome goes AWOL in men as they head toward their golden years. The study of two hundred thousand men in the UK found that the proportion of participants with significant loss of Y rose from 3 percent at age forty to 44 percent among those at age seventy. According to some scientists, it is the most common mutation of *any kind* to arise after birth in men. Loss of Y can also be extraordinarily pervasive within the body: A small study found that one man was missing this chromosome in almost 90 percent of his sampled blood cells.

How do cells missing a chromosome become so abundant? When Jacobs and her coauthors published their findings decades ago, they put forward a few theories. One possibility, they wrote, was that dividing cells in older people are more likely to make mistakes when sorting their chromosomes. Or, perhaps, cells with an abnormal chromosome count somehow stick around. When it comes to loss of Y specifically, scientists have speculated that the Y chromosome may carry a growth-suppressing gene, so jettisoning the chromosome would free a cell from that constraint. There is also the possibility that cells undergoing some other change that gives them an advantage might *coincidentally* drop their Y chromosome in the process.

Although less is known about loss of X—in which one of the two X chromosomes that women carry starts disappearing from their cells—there's an inkling that it's common, too. Jacobs is one of the few scientists to publish about this. In 2007, she coauthored a study of almost twenty thousand cells from more than 650 girls and women that detailed a "highly significant" relationship between the X chromosome's disappearance and aging. Whereas around 0.07 percent of girls younger than age sixteen showed X chromosome loss, the number climbed to 7.3 percent of women by age sixty-five. (Sadly, the number of published papers related to this phenomenon in women remains paltry.)

There's still plenty more mystery to unravel about loss of Y and the disappearance of the X chromosome. But one thing is becoming clear: Even if people seem perfectly healthy as they grow older, there can be populations of abnormal cells taking hold in their bodies.

. . .

Compared with mysterious genetic conditions, heart disease can seem comparatively banal. It's unfortunately abundantly common. If you are a human being—and if you are reading this that probably describes you well—the ailment most likely to kill you is heart disease. Heart disease claims the life of about one out of every six people on the planet. It is, according to the World Health Organization, "the world's biggest killer," and in many places around the globe it is on the rise. Even in the US, where recent strides have been made against cardiovascular illnesses, heart disease has been the leading cause of death for the past century.

Despite the fact that heart disease looms so large, it was nowhere on Siddhartha "Sidd" Jaiswal's research agenda when he was a newly minted doctor in California. Jaiswal was involved in the treatment of patients with acute myeloid leukemia, a rapid and deadly cancer that starts in the bone marrow and can spread to the rest of the body. He had immense compassion for these patients, but he knew that their prognosis was poor: Only around one in four will survive five years after their diagnosis. This dismal situation gnawed at Jaiswal.

In the early 2010s, he moved across the country to Boston for a postdoctoral fellowship at Harvard Medical School and started reading everything he could about how leukemia takes root in the body. He came across a paper that grabbed his attention. Researchers in Montreal had found that as some women age, their blood cells skew toward one of their two X chromosomes, suggesting a sort of competition in which one population of cells was winning, similar to what Patricia Jacobs and others had seen before with loss of Y. The Canadian group found something else: The elderly women with this skewing often had some cells with mutations in a gene called *TET2*. *TET2* had been associated with blood cancer.

When Jaiswal read the report, he was gripped. He wanted to know if having mutations in genes such as *TET2* was a risk factor for developing leukemia. If this was true, it would provide an opportunity to identify people prone to the disease before it took hold and progressed beyond the reach of treatment.

Jaiswal joined a lab at the Broad Institute in Cambridge, Massachusetts, with a tremendous trove of DNA sequencing data and got going. When all was said and done, he and his collaborators had genetic data from seventeen thousand people from countries stretching from Sweden to Korea. Their analysis confirmed that as people got older their bodies were more likely to harbor populations of blood cells with mutations in genes such as *TET2*. And the data also confirmed Jaiswal's hunch: Individuals with those mutated cells had a tenfold increased risk of developing blood cancer, such as leukemia. But he was stunned to find many more people than expected who had these abnormal cells in their blood, and they also had a heightened risk of death even when they were cancer-free. Jaiswal wanted to dig deeper. "I started to really think, these people have these massive [numbers of] clones in their blood that are mutated," Jaiswal recalls. "That's got to have some consequences besides just leukemia."

Jaiswal was curious whether the presence of mutant blood cells was linked to cardiovascular illness. Unbeknownst to him at the time, a father-and-son research duo—Earl Benditt and John Benditt—had found evidence decades earlier that blood vessel plaques in women were clonal outgrowths from a single cell. The plaques were either solely or predominantly made up of cells with the same active X chromosome from the two X chromosomes that these women carried, rather than the expected fifty-fifty mix as in the artery walls. But Jaiswal's collaborators in Cambridge were skeptical when he told them he planned to see whether mutant clones were linked to cardiovascular issues. He went ahead with it anyway, and the results were striking: People with mutant populations of blood cells also had a doubled risk of coronary heart disease and a more than doubled risk of stroke. "That's when things really got a little interesting," Jaiswal told me.

The study from Jaiswal and his collaborators appeared in *The New England Journal of Medicine* in 2014—and it made huge waves in the years that followed. It helped cement the idea that the mutant cells that arise within us can cause more than cancer or rare disease. The results made clear that mutant cellular clones in the blood can contribute to the biggest killer on Earth, heart disease.

Jaiswal and some of his coauthors called the condition "clonal hematopoiesis of indeterminate potential," or CHIP for short. Hematopoiesis refers to the body's process for making blood cells, and "indeterminate potential" acknowledges the haziness of the condition. Although the mutant clones elevate an individual's risk of a growing list of diseases, they don't seal a person's fate. There are plenty of folks walking around with CHIP who have not experienced a bad health outcome. In fact, it is stunningly common. It's now estimated that around 10 to 20 percent of people in their seventies have it. (Smaller populations of mutant cells with CHIP-like mutations are present in most folks by age fifty, although the number of such cells in these younger individuals is usually too few to meet the definition of the condition.)

Clearly, aging is a big risk factor for CHIP. But finding other possible determinants is a hot area of research. Indirect evidence suggests that some genetic or environmental factors are at play. There are also hints that obesity might exacerbate CHIP. Even fragmented sleep might perturb the balance of cells in the blood.

Scientists have been so fascinated by CHIP that they have studied how it is influenced by time spent in outer space. An analysis of stored blood samples from fourteen astronauts who flew space shuttle missions found that more than a third of them had blood cells with a mutation in *DNMT3A*—one of the genes linked to CHIP by Jaiswal and others. The frequency of mutant blood cells in the astronauts' samples was less than 1 percent—which is below the 2 percent threshold to be considered CHIP. As such, none of the astronauts would officially be diagnosed with the condition. Still, the researchers behind the study said that the findings were striking given the relatively young age and good health of the astronauts. They speculated that some of the CHIP-related mutations might have been precipitated by exposure to space radiation.

There are some hints that CHIP doesn't even have to originate in your body for your health to be negatively affected by it: A number of studies suggest that people who receive blood from donors with CHIP may face an increased risk of developing leukemia.

■ ■ ■

Now that CHIP has been discovered, the hunt is on to understand it better. Like Lucio Luzzatto, who studied PNH, Sidd Jaiswal has become preoccupied with how mutant cellular clones gain a foothold. "I think the key thing to remember is that in order to get CHIP, you have to have a mutation that confers a clonal advantage," says Jaiswal, now a pathologist at his alma mater, Stanford.

Blood is a ripe setting for a takeover of errant clones because of its high turnover. There are around twenty-five trillion red blood cells coursing through a person's body, each one dutifully helping to deliver vital oxygen to tissues. They are hard workers, but they are also relatively short-lived: Each will last around 120 days before getting recycled for its parts. To compensate for this loss, every second or so, your body creates 2.4 million new red blood cells. These ongoing births trace back to deep in your bone marrow, where stem cells and blood progenitor cells live. But over time, DNA changes can pile up there. One research team calculated that by age fifty, the average person has accumulated around five mutations in coding genes within each of their blood progenitor cells. "This sets the stage for a robust Darwinian selection of mutations that provide a competitive advantage to the mutant [progenitor cell] by promoting its self-renewal, proliferation, or survival," geneticists José Fuster and Kenneth Walsh wrote. Some mutant progenitor cells might pick up genetic changes allowing them to either replicate faster, churn out new mutant blood cells more rapidly, or simply stick around longer. Ultimately, when cells with certain mutations have reached large enough numbers in the blood, a person is officially diagnosed with CHIP.

The forces of evolution "apply not only to individuals in a species but also to cells within the body," according to another geneticist duo, Carl Anderson and Sigurgeir Olafsson of the Wellcome Sanger Institute in Britain. However, they argue that whereas species evolution is often discussed as being influenced by negative selection—a process by which detrimental variants are purified out over time—the dynamic within bodily tissues is dominated by positive selection. The latter describes a situation where a beneficial genetic variant confers an advantage that is selected for and spreads (in this case, within a population of cells). Anderson and Olafsson add that the body is a shifting landscape, especially when a person is ill. If a disease alters a person's tissues, for example, mutations that were once neutral might suddenly give the cells that carry them an advantage.

So how do CHIP mutations—which are often in genes associated with cancer—heighten the risk of heart and blood vessel disease? According to Jaiswal, the answer has to do with inflammation. When he and his colleagues took mouse cells with perturbed *TET2* and exposed them to cholesterol, the cells responded with immune signals reminiscent of how they react when they encounter a bacterial toxin. Inside the human body, this kind of unusual response may result in increased inflammation, which is known to harden blood vessels and cause heart damage.

Jaiswal suspects that inflammation might also be at the root of a possible link that his team identified between CHIP and heart arrhythmias. His theory is that genetically divergent clones cause some sort of immune disturbance that leads to scarring in the heart, which in turn disrupts the regular electrical pulses inside it.

Given all this, some scientists have floated the idea that certain people with CHIP should be offered anti-inflammatory medications as a protective measure against heart attacks and stroke. One such scientist is Michel Goldman. Goldman is an immunologist who previously oversaw a $2 billion research endeavor in Europe called the Innovative Medicines Initiative and is founding president of a university hub in Belgium aimed at assisting in healthcare projects that include drug regulation.

The matter of CHIP is personal for Goldman. At the end of the summer of 2021, he developed a rare form of cancer, and genetic sequencing of a sample of his bone marrow cells found that more than 72 percent of them harbored mutations in *TET2* and 82 percent harbored mutations in *DNMT3A*. These were genes that Jaiswal and his collaborators had tied to CHIP. There was an evolutionary battle among Goldman's cells, and the ones with the mutations appeared to be gaining an edge. It was an ominous finding. Goldman went through six brutal rounds of initial chemotherapy, after which his cancer was declared to be minimized and well controlled. But a year and a half after his initial cancer diagnosis, he suffered a heart attack. Although a stent saved his life, he wonders if CHIP had been a contributing factor to his heart attack, and—if so—whether anti-inflammatory medicines would have averted that cardiac crisis.

Meanwhile, more connections have been found between inflammation and acquired genetic changes in the blood. Less than a decade after Jaiswal and his team linked cellular clones to cardiovascular disease, another group began reporting on a previously unheard-of inflammatory illness—and the leading culprit appeared to be mutant blood cells.

■ ■ ■

Chuck Stoner's first signs of sickness turned up in 2009. At the time he was forty-five years old, working as a chemist in the army and living in the same Maryland house he had shared with his wife and family for two decades. Stoner had a regular exercise routine, but he started experiencing weird abdominal pain when he would lift weights. Later, when he played in his community softball league, he had recurrent pulled muscles. The repetitive injuries mystified him. As the year progressed, the symptoms became stranger. Each afternoon he felt he was coming down with a cold or flu, but he would miraculously rebound. "I'd feel fine in the morning, I'd go to work and then I'd get that feeling like, 'Oh, man, I'm getting sick,'" Stoner recalls. "And then the next morning I'd feel fine again."

It wasn't all in his head. Stoner's daily malaise was increasingly accompanied by real fevers—some of which would be as high as 103 degrees Fahrenheit. His doctor sent him to a specialist, who prescribed him steroids, which seemed to give Stoner relief. But long-term use of steroids can cause side effects such as diabetes and osteoporosis. So when Stoner had flare-ups, he did his best to cope using over-the-counter painkillers. He would take some Advil, lie on the couch, and "sweat it out for a couple hours" until he felt he could function again.

The next decade brought more mysterious symptoms. He developed lumps under his skin from time to time, and "nasty-looking" rashes. His baffled doctors eventually referred him to the US National Institutes of Health Clinical Center, which runs a program to help patients with problems like chronic, inexplicable fevers. In January 2019, he went to the clinic for a check on his heart. "I went down there for a heart MRI and they said, 'You're not leaving,' because my heart function was like 29 percent," Stoner says. "They thought I was going to drop over." He had inflammation of his heart tissue and ended up in the hospital for more than two weeks. That year he retired from his civilian job within the army on disability. The mystery illness was starting to take a deep toll on his life. Thankfully, an answer to the question of what ailed him—and its surprising link to evolutionary principles—was around the corner.

A young doctor named David Beck had recently joined the NIH. Beck recalls that "we had so many patients that we couldn't find the answer for"—patients such as Chuck Stoner. Previous research had often focused on very small groups. "A lot of how genetics works is we go case by case looking to see if we can find a causative variant in a family," Beck explains. "But I was kind of keen to stop doing that and to leverage all of the cases that we had within the clinic so that we could see if there were shared variants." Beck and his colleagues decided to study a large population of patients, including Stoner, to see what mutations they had in common.

Stoner's blood, along with that of many other patients at the clinic and some from hospitals in England, was sent for DNA analysis. A project this massive required careful coordination. Beck and the head of the lab where he worked, Dan Kastner, collaborated with more than fifty scientists to access and crunch all the data. Ultimately, the study turned up a culprit causing Stoner and two dozen other men to have symptoms such as recurrent fevers and torturous inflammation. The blood cells of these men had acquired mutations in a gene called *UBA1*, which produces an important enzyme that helps cells dispose of their waste. The faulty garbage system of Stoner's and other patients' cells was somehow causing their illness.

When Beck saw the results, he was struck. "It was like an 'aha moment,'" he recalls. Finally, he could explain the genetic problem behind Stoner's sickness. Just like with blood conditions such as PNH, loss of Y, and CHIP, mutant cells had taken over.

In late 2020, the team published a paper describing the brand-new and often fatal disease, which they named "the VEXAS syndrome." The name is actually an acronym for a string of words that doesn't easily roll off the tongue. It is short for "vacuoles, E1 enzyme, X-linked, autoinflammatory, somatic" syndrome.

Beck adds that many of the people in his study who had been suffering from VEXAS had been reluctant to describe all their symptoms before they knew they had a defined disease. That all changed when Beck told them they had VEXAS and asked them about any manifestations of the illness, such as debilitating fatigue. "When we came back and we asked, 'Do you experience this?' they said, 'Yes, but I never told anyone.'"

Scientists have estimated that about one in every twenty thousand or thirty thousand individuals over the age of fifty have VEXAS in the US alone. "Basically every hospital system is going to have a few people with the condition," Beck says. But a majority of those patients are not being diagnosed. "I think we've found something that hasn't quite hit the textbooks," he explains, "so patients aren't really being recognized with the disease yet."

Back in Maryland, Stoner is coping as best as he can with VEXAS, despite still having to take medicine to stave off flare-ups. He remains hopeful that there will be new drugs for the disease in the future. In some ways, he is one of the lucky ones. By the time scientists were able to publish the data from him and the other twenty-four men they initially studied, ten of them had already died from complications related to the disease, such as respiratory failure and severe anemia. The terrible toll of VEXAS that was hidden for so long is now coming to the surface. It's a stark example of why we need to pay more attention to the mutations we pick up as we age.

The question that now remains is how cells with mutant *UBA1* edge out their neighbors to cause VEXAS. "We think that the mutant cells aren't just having a survival advantage, but they're also creating a sur-vival disadvantage for the neighbors," speculates Beck, who now has his own lab at New York University's Grossman School of Medicine.

One person who was impressed by the VEXAS findings was Sidd Jaiswal. He said that the rise of blood cells with the *UBA1* mutation is a prime example of the kinds of disorders with Darwinian dynamics that he and others are exploring. "Here you have a mutation where if you acquire it, you have this very severe autoinflammatory disorder," Jaiswal says, "so it's a really great example of how extreme the manifes-tation of the genetics can be."

The good news is that scientists are no longer only documenting the rise of mutant blood cells. They have plans to find the weak spots in these rogue clones—and then stop them in their tracks.

◼ ◼ ◼

A lot has changed in Sidd Jaiswal's life since he published his 2014 paper describing the blood condition that would be known as CHIP. He got married, bought a house, and became a dad. Now in his mid-forties, his thick black hair and trim beard have acquired a salt-and-pepper color. "I can see the effects of aging on myself," he jokes, though he also notes that he has not yet tested himself for CHIP.

Jaiswal is now part of an effort at Stanford to help patients who have a recent CHIP diagnosis. People with the condition often have questions about their future risk of cancer and heart problems. His colleague Tian Yi Zhang spearheaded the launch of a CHIP clinic within the Stanford system several years ago. At the start, she saw only about a dozen new patients a year, but within a few years that number has approached around a dozen each *month*. The clinic is still expanding. "We're all in the 'If you build it, they will come' stage," Zhang says.

More and more hospitals are launching similar clinics with the expectation that people with CHIP will need counseling and advice. CHIP clinics have cropped up in recent years at venerable places like the Dana-Farber Cancer Institute in Boston, the Memorial Sloan Kettering Cancer Center in Manhattan, and the Vanderbilt University Medical Center in Nashville. The Dana-Farber CHIP clinic is known as the Center for Early Detection and Interception of Blood Cancers and aims to help "individuals who are at high risk for developing blood cancers due to precursor conditions."

At all these clinics, patients are given support for a condition that brings a lot of uncertainty. CHIP is associated with cancer and heart problems, and it takes a lot to wrap one's head around such a diverse range of risks. The patients enter a phase of "watchful waiting" to ensure that they stay one step ahead of any health consequences from CHIP. It can be tremendously anxiety inducing. They may be referred to a cardiology expert, a social worker or psychiatrist, or even a financial planner to help them prepare for the possibility of a future illness arising from their CHIP condition.

Zhang meets with patients once a week at Stanford's CHIP clinic. "It's physically at the Stanford Cancer Center, which is slightly weird for some folks because they don't have cancer," she says. Zhang and her team test each patient once or twice a year to see whether the mutant cells in their blood have started to grow out of control. For patients who have acquired mutations linked to a high risk of cancer, she also recommends a bone marrow biopsy to make sure no malignancies have developed. If she catches a blood cancer in its nascent state, she recommends early interventions, such as a bone marrow transplant.

Zhang says that some patients prefer not to learn too much about how their mutant blood cells are behaving, but "other people will say, 'I want to know. I want a plan.'"

Perhaps one of the most important functions of CHIP clinics is to offer useful guidance when patients are choosing what therapies to take. For example, certain kinds of chemotherapy can actually push mutant clones in the blood to higher levels, elevating the risk of secondary cancers. If patients are diagnosed with CHIP, they can be counseled on whether to avoid those specific treatments.

Jaiswal sees a future where CHIP patients receive much more than counseling and guidance. He envisions a time when their condition can be dramatically curbed. "To be honest, my main feeling is, 'Why do we not have a drug yet to help these people?'" he says. "And that's really what's kind of motivating this next phase of my career." Ideally, a treatment could tamp down the populations of mutant clones in these patients' blood.

After a decade of searching, he has identified a potential target for such a treatment. He and his collaborators hunted in the genomes of more than five thousand people with CHIP and found that those with a faulty variant of the *TCL1A* gene had a slower growth of mutant clones in their blood. This implies that certain CHIP mutations need a working copy of *TCL1A* in order to wreak havoc.

Jaiswal imagines that drugs designed to target and interfere with *TCL1A* might help treat—and perhaps even reverse—disorders such as CHIP. Initially the medication might be given to people whose CHIP has already progressed to the point where full-blown blood cancer has developed, but in the future, it might be given to those who simply have growing populations of mutant clones in their blood. The latter approach—prescribing a therapy to stop cancer from ever evolving— would be so novel that it would tread into uncharted territory. It's not something that the US Food and Drug Administration (FDA) is used to approving. "If you're talking about giving somebody a drug for life to prevent them from getting cancer, that's a different ball game and not a development path that's been laid out by regulatory agencies like the FDA," Jaiswal explains.

■ ■ ■

The idea of treating diseases caused by mutant blood cells is not just a pipe dream. Take PNH: It was one of the first such diseases to be discovered, and one of the first for which working therapies now exist.

You might recall from earlier that in PNH a spontaneous genetic error causes blood cells to begin making a faulty version of the normally protective PIG-A protein. These mutant cells somehow start crowding out their regular counterparts. But later, the disease advances to a stage where red blood cell counts drop overall. That's because the mutant cells, which lack a working version of PIG-A, start getting destroyed by the immune system.

If left untreated, PNH can cause life-threatening blood clots. Most people with PNH used to live only ten to twenty years after their diagnosis. Half a century ago, doctors began trying bone marrow transplants to reset the blood-making system in PNH patients. But bone marrow transplants are inaccessible and too costly for many people around the world—not to mention that the procedure also carries a significant risk of death.

Then came a breakthrough.

Scientists developed a drug called eculizumab that could swoop in and shield cells with defective PIG-A from demolition. In 2007, the FDA approved the medication. People with PNH receiving eculizumab and similar drugs avert the risk of clots, and it's believed they might have a normal life expectancy.

The power of these drugs is real. Take for example the case of a twenty-year-old college athlete who had multiple episodes of vomiting and abdominal pain along with an inflamed vein at an IV-line placement site during an ER visit, the latter of which suggested blood-clotting issues. The student recalled having had multiple episodes of dark urine. His condition was followed by Daria Babushok, a hematologist at the University of Pennsylvania's Perelman School of Medicine. She and her colleagues found evidence of PNH and—thanks to advances in diagnostics—were able to track the expanding populations of his mutant cells over several years. The athlete was started on eculizumab, and although he remains at risk of blood clots, his PNH symptoms disappeared.

There's a notable catch to eculizumab. It is inordinately expensive—the medication can cost more than $500,000 a year. That makes it one of the most expensive drugs on the entire planet. For this reason, Luzzatto is reserved about the success of eculizumab and similar PNH treatments that followed. "I am not so triumphant," he told me. "At least two-thirds or three-quarters of patients in the world have no access given the exorbitant cost of these drugs."

In the United States, each year several hundred people get a diagnosis of PNH, and there might be many others who go undiagnosed. Determining whether someone has PNH is not always straightforward. Thirty years after Luzzatto and his colleagues in Nigeria made the discovery of PNH clones, he was part of a team that detected these rogue cells in healthy people. They analyzed blood samples from nine normal volunteers and found that *all* of them had PNH clones in their system, with an average frequency of around twenty-two per one million sampled blood cells. At the time, the finding was received with some skepticism, but in recent decades, other studies have also found PNH blood cell clones in healthy people who do not have the disease.

This has huge implications. If you pause to think about it, this means that in some of us a small population of rogue cells is poised to edge out their normal counterparts if the conditions tip in their favor. This only rarely precipitates PNH symptoms, so it's not something that should stir anxiety. The mutant cells are sometimes merely transient—at some later point disappearing forever. But if a doctor finds PNH clones circulating in a patient's blood, when can they be certain that the individual truly has PNH? The question is particularly vexing because some of the symptoms of PNH overlap with those of other diseases. Luzzatto stresses that a person truly has the disease only if they have symptoms such as hemoglobin in their urine, which happens only if the mutant cells have grown large in number. He explains that if, for example, less than 0.1 percent of a patient's blood cells have the PNH mutation, it is unlikely that they actually have the disease.

In recent decades, the scientific community has become vastly more receptive to the notion that cells with genetic changes can circulate in the human body. It's becoming evident that this is a very common phenomenon that happens in particular as we age. "I've been telling students for many, many years, 'Being older than you, I have accumulated a lot more somatic mutations,'" Luzzatto says. There's a greater appreciation that our cells are not genetically identical, and that some of them mutate in consequential ways that can give us diseases besides cancer.

The conversation about reining in the mutant cells—the populations of "clones" that arise in our bodies over time—can quickly turn philosophical. "Being a multicellular organism, the trade-off of that is you have these competing clones in your body and you can't really stop that process from happening. It's simply a by-product of being alive," Jaiswal says. "But maybe what we'll be able to do in the future is to steer the evolution in certain directions, to maybe have clones circulating in our body that are more benign or even protective against certain diseases of aging."

The idea of overcoming bad mutant cells with good mutant cells that Jaiswal describes is not as far-fetched as it might seem. In fact, it happens naturally in some individuals with rare disorders thanks to a genetic chance event (more on this in chapter 6). In these people, "good" mutant cells serendipitously appear. When the helpful cells arrive on the scene, they come to the rescue and stamp out the preexisting disease. It's a magnificent example of how the body can experience healing genetic changes during a lifetime.

Scientists are also curious about possible similarities between different mutant blood cell conditions. A study of men who had undergone a surgery to remove plaque buildup in the arteries that deliver blood to the brain found that one out of six had loss of Y. Those with the genetic anomaly had a doubled risk of major cardiovascular events after the procedure. This meant that loss of Y, like CHIP, was also linked to cardiovascular woes. Researchers engineered male mice to be deficient in the Y chromosome in their blood cells and saw that the rodents were prone to heart failure.

The similarities between loss of Y and CHIP compelled some members behind those studies to investigate whether these two conditions were simply "two sides of the same coin." They looked at the blood cells of almost two dozen men and found the "striking observation" that nine of the twelve who had loss of Y also had gene mutations associated with CHIP, whereas only one of the ten men without loss of Y had CHIP. Other teams are chasing an answer about a possible connection between these conditions. In 2023, UK researchers parsed genetic data and found that loss of Y was linked to mutations in *TET2*—one of the most prominently mutated genes in CHIP. Additionally, a variant of the *TCL1A* gene that seems to decrease the risk of CHIP is also associated with a reduced risk of loss of Y.

The convergence of these findings is thought-provoking. We clearly still have a lot to learn about how mutant blood cells become prevalent in many people as they age. Conditions such as PNH, loss of Y, CHIP, and VEXAS—all fueled by acquired genetic changes—challenge our traditional framing of genetic disease. And if we look beyond them, there are even more surprises.

5.

The Heart of the Matter

I n 2013, Sici Tsoi was expecting her third child. Everything seemed fine until she went for a routine checkup in her thirtieth week of pregnancy and her obstetrician noticed a troubling irregularity in the baby's heartbeat. The potential diagnosis had an immediate impact on Tsoi's pregnancy. She had to go to the hospital regularly so that the medical team could use a special ultrasound test to track her baby's developing heart. These checkups continued until her thirty-sixth week of pregnancy. Tsoi came in for a fetal ultrasound, but this time doctors didn't let her return home. They had spotted a concerning fluid buildup around her baby's heart, which suggested the organ might be failing. The consensus was that an emergency cesarean section was necessary right away to protect Tsoi's baby.

The birth of Tsoi's daughter, Astrea, at the Lucile Packard Children's Hospital at Stanford happened quickly. Tsoi's husband, Edison Li, barely got there in time. "I went inside the delivery room at 5:41 p.m. and Astrea came out at 5:45 p.m.," he says. Still, neither Tsoi or Li saw their new daughter in the delivery room after she was born. "They immediately took her away to do a checkup to find out what was the issue with her heart," Li explains. Astrea had been whisked away to the neonatal intensive care unit. There, the situation only became more intense: Astrea went into cardiac arrest.

The medical team suspected that Astrea had a condition known as long QT syndrome, and worryingly, it looked like it was a severe form of the disease. Congenital long QT syndrome is the result of a genetic defect that prevents cells from efficiently recuperating the electrical charge that keeps the heart running like clockwork. This life-threatening disorder sends the heart out of whack. Its exact prevalence is tough to pin down, but by some estimates it causes around one thousand deaths each year in the United States alone, mostly in children and young adults.

Tsoi and Li were told that their newborn daughter would need to undergo an operation immediately to implant a cardioverter défibrillator device in her heart. The device would reset the rhythm of her heartbeat if it ever fell out of sync. Tsoi and Li consented to the procedure, and the surgical team implanted the device successfully.

Before the family could go home, however, doctors decided they would do one more step. They wanted to look for the actual mutation that Astrea had so that they could confirm her suspected long QT syndrome diagnosis. The Stanford University–affiliated hospital was trying out new DNA analysis tools that gave them the ability to sequence the protein-coding genes in people's cells. Pediatric cardiologist James Priest, who was in his mid-thirties at the time, wanted to try this with Astrea's cells. "It was pretty likely that she had an inherited arrhythmia syndrome," Priest recalled to me. "My postdoctoral adviser at the time, Euan Ashley, was like, 'Let's try to sequence her as quickly as we can to get the genetic diagnosis.'" When Astrea was just three days old, her blood was drawn for the genetic analysis that promised to give a clear-cut answer.

Initially, Priest found what he was looking for. A DNA readout showed that Astrea possessed a mutation in a gene called *SCN5A*, which encodes channels in heart cells that regulate the flow of sodium—and therefore influence the flow of electrical currents. Errors in the *SCN5A* gene are known to cause long QT syndrome, so Priest thought it was an open-and-shut case. It was also helpful to know about the *SCN5A* mutation because it guided the decision about which medication to give Astrea (certain drugs for long QT work better on sodium channels than others). But then doubt started to creep in: When Priest did a follow-up test on a small sample of her white blood cells, he didn't find the mutation at all.

Priest was baffled. Long QT is usually inherited, so he then tested Tsoi and Li for the *SCN5A* mutation, partly for their own benefit: "If one of the parents has this genetic difference, they could be at profound risk of having life-threatening arrhythmias," Priest explains. But once again, the tests turned up no mutation. There was a chance that the mutation had arisen in the sperm or egg cell that made Astrea, but if so, it would have been in all her cells. That was not the case—it wasn't in her urine or hair follicles. Something wasn't adding up. "The only explanation was that there was something even more complicated going on," Priest says.

Priest and his colleagues began to suspect that Astrea's body contained a mosaic of genetically different cells. The thinking was that when she was a developing embryo, a mutation popped up in *SCN5A* inside one of her cells that then multiplied to become many cells within her body. To get a better sense of whether this hypothesis was right, they connected with fellow Stanford researcher Stephen Quake.

Quake had helped pioneer new tools that enable the genetic sequencing of individual cells, and now he would take Astrea's blood cells and analyze their DNA one by one. When he did, it suggested that the vast majority of her blood cells had a normal copy of *SCN5A*. Only three out of the thirty-six blood cells analyzed carried the mutation in that gene. The results revealed that Astrea had mosaic blood; the mutation for long QT syndrome must have cropped up during her fetal development. The Stanford team felt they finally had the much-needed answer to the mysterious case of Astrea's heart arrhythmia. They wrote up a paper describing the results for submission to a scientific journal.

The group learned, however, that other scientists in the field were not fully persuaded. Definitively proving the presence and the effects of a mutation on the heart can be tricky in a living person because the organ is sensitive and deeply buried. Quake had analyzed cells from Astrea's urine and inner cheek, which—just like her blood—had similarly suggested that small percentages of her cells from those tissues had the *SCN5A* mutation. But these cells, too, offered only indirect evidence that she had long QT. "One of the main criticisms that we got was, 'Well, you don't know that this mutation is in the heart,'" Priest says.

After more than a month at the hospital, Astrea came home, but the genetic makeup of her heart cells remained merely an inference.

■ ■ ■

Some cases of mosaic genetic conditions are easier to spot than others. There are even ones that you can see with the naked eye. It's no surprise that those visually striking examples were among the first mosaic conditions to be described in medicine. Take, for example, a condition presented back in 1901 during a scientific meeting in the city of Breslau. The conference had attracted a German dermatologist named Alfred Blaschko, who shared his observations of curious birthmarks and skin lesions on 140 patients. In these people, bands of darker and lighter skin appeared along similar lines of the body. The marks would form a series of contours in a V-like dip along the upper spine, an S shape along the abdomen, and an inverted U across each breast. Blaschko believed that the patterns resulted from something that happened during embryonic development.

In the years that followed, these enigmatic bands of differently pigmented skin began to be described as following the "lines of Blaschko." Other doctors found even more unusual examples in patients. A Soviet surgeon named Moisey Zlotnikov documented a case of a woman who had bands of skin of various colors. The woman had been mocked as a devil when she was a child. Some of her skin was deep brown, whereas other sections were partly crimson or a light brown hue that dermatologists refer to as "café au lait." The bands of color appeared on her skin from her head on down—but only on her left side. Her right side appeared completely normal; it was totally uniform in color.

Zlotnikov was living in an era when scientists had become increasingly taken with the idea that some sort of genetic material existed within cells and guided their fate—and that it could change. He suggested that the woman's skin condition had resulted from an early embryonic *mutation*.

Unfortunately, Zlotnikov published his case report in 1945, as World War II came to an end, and in all the turmoil of that time, the report went unnoticed. His idea remained in obscurity. Later, in the 1970s, scientists rediscovered the lines of Blaschko. Among them was a German dermatologist named Rudolf Happle, who became one of the researchers to hypothesize that the lines of Blaschko resulted when genetic changes occurred in *some* skin-forming cells during early embryonic development. Those abnormal cells would grow along developmental lines to form areas of differently pigmented skin that contrasted against normal sections. Thanks to modern genetic technology, in the 1990s scientists were able to confirm the theory. Pigmentation marks that follow the lines of Blaschko now represent a prime—and visually striking—example of how humans can be genetic mosaics. Such skin patterns illustrate that humans pick up new mutations during their development and growth.

A condition that is more familiar, cancer, is also a mosaic disorder. In cancer, the mutated DNA of malignant cells contrasts with the normal genetic makeup of their benign counterparts. And when a tumor grows on or near the surface of the skin, it can—just like the skin marks that follow the lines of Blaschko—be seen by the naked eye. The overt manifestations of cancer are perhaps why it was the first illness that scientists linked to acquired mutations.

Sometimes the turbulence caused by mutated cells is hidden beneath the surface. The conditions described in the previous chapter—PNH, loss of Y, CHIP, and VEXAS—are instances in which the blood contains a mosaic of genetically dissimilar cells. And although they might not be visible to the naked eye, they are relatively easy to diagnose. Unlike with mosaic conditions that affect deeply situated internal organs like the heart, testing for those that affect circulating cells can require as little as a simple blood draw.

Daria Babushok, a hematologist who treats PNH, says that mutant cells are probably competing against normal counterparts in many places within the body, even if we can't always document it. "Basically, you have this microevolution that's happening in all of us," she says. "It's not just in PNH—it is happening in different organs. It's just that blood is more accessible. You can take a blood test and you see it."

Babushok's words ring true. Slowly but surely, the profound health influence of new mutations is being discovered in more organs throughout the body.

■ ■ ■

The brain condition known as hemimegalencephaly is difficult to pronounce, but its manifestation is simple to see on a brain scan. In many cases it takes only a quick glance at the MRI readout to know that something is terribly amiss. In hemimegalencephaly, one half of the brain is larger than the other. Instead of a symmetrical image, the areas of white and gray matter appear disorganized and out of proportion on the scan. The condition can cause epilepsy and intellectual disability.

In 2006, a team at the UCLA Mattel Children's Hospital published a paper on the disease that included a pair of identical twins, one of whom had hemimegalencephaly. The fact that only one twin was affected suggested to researchers that the disease might have to do with a mutation that occurred during development in the womb rather than one that was inherited.

Across the country, Christopher Walsh had recently become head of the Division of Genetics and Genomics at the Boston Children's Hospital. There, he had a chance conversation with Annapurna Poduri, director of the hospital's epilepsy genetics program. Poduri thought that the causes of hemimegalencephaly deserved a closer look. The nature of hemimegalencephaly on brain scans was a big clue that it did not behave like neurological diseases that are passed down the family tree. If an inherited mutation caused hemimegalencephaly, then one would expect all of the brain to be affected the same. But that was not the case. The condition "looked anything but inherited, because only half of the brain was abnormal, and the other half was normal," Walsh recalled years later.

Brain tissue—like heart tissue—is difficult and dangerous to access for genetic sampling. There's a skull protecting it, and you can't simply grab a bunch of brain cells and hope that the organ will function normally, just to list a couple of the *many* challenges. However, there's a situation that can arise in hemimegalencephaly that creates an exceptional situation: In some cases, the seizures it causes can be so resistant to treatment that doctors must take out parts of the brain or—in an extreme measure—remove half of it completely. Children who are severely affected by the disorder have no chance for a normal life without these drastic interventions.

Walsh and his colleagues saw the discarded brain tissue that is a byproduct of these procedures as an opportunity to get a glimpse into the genetic changes underlying hemimegalencephaly. In the excised tissue, Walsh and his collaborators found mutations linked to genes that control the growth of neurons. What was striking was that those mutations were not always found in the children's blood or other parts of their brains. And even within affected brain areas, the variants sometimes appeared in only around 8 percent to 35 percent of cells. This low frequency supported an idea that the patients had acquired the DNA changes in parts of their brains spontaneously during development. Walsh and his teammates speculated that the timing of the acquired mutations also influenced the size and severity of each patient's hemimegalencephaly. They published their findings in April 2012.

Other doctors and scientists had been similarly spurred to investigate what was going on in hemimegalencephaly. They included Joseph Gleeson, a professor of neurosciences and pediatrics at UC San Diego School of Medicine. Just a couple of months after the report from the

Massachusetts team came out, Gleeson and his collaborators published findings from their own hemimegalencephaly study. They had studied samples from twenty patients and found that a third of them had spontaneous—noninherited—mutations in similar genes. Both the Massachusetts and California teams learned that the spontaneous mutations behind hemimegalencephaly often affected genes involved in helping cells coordinate with one another. The coordination in question relies on a molecule called mTOR.

There was a growing consensus of the mTOR pathway's importance. A paper published alongside the one from Gleeson and his colleagues, from a third ensemble of scientists, also found spontaneous DNA errors in this pathway in cases involving asymmetric brain enlargement and hemimegalencephaly. Genetic changes in the mTOR pathway also turned up in a hard-to-treat form of epilepsy called focal cortical dysplasia.

"The Gleeson and Walsh papers showed that somatic mutations in the mTOR pathway could lead to cortical overgrowth and epilepsy," says Mike McConnell, a neuroscientist and founder of Rare Mosaic Scientific Consulting, who has previously collaborated with them. "These were fundamental links between brain somatic mosaicism and brain disease. In other words, somatic mutations aren't just something interesting that happens, they could have profound consequences in some individuals."

Finally, there was credence to the notion that spontaneous—or so-called *de novo*—mutations in certain genes could be at the root of some very serious brain diseases. The evidence also added weight to the idea that cells in the brain aren't all 100 percent genetically identical.

What was perhaps most exciting was that some of the new findings could point to new treatments. Not only was it becoming indisputable that these spontaneous genetic mutations were happening, there was a growing appreciation that understanding them could help drug developers know which molecular pathways to target.

Take, for example, the insights into how somatic mutations in the mTOR pathway can lead to disorders such as focal cortical dysplasia. Drugs called mTOR inhibitors that affect these molecular signals are being explored as a treatment for certain forms of epilepsy.

Jeong Ho Lee, who was a postdoctoral researcher in Gleeson's lab and the lead author on the group's 2012 paper showing somatic mutations in hemimegalencephaly, has been pursuing this goal. Lee, a scientist at the Korea Advanced Institute of Science and Technology, now serves as chief scientific officer for a company called Sovargen, which is pursuing epilepsy treatments that target the mTOR pathway. The company entered a licensing agreement to test the drug paxalisib, which was initially developed for cancer. Paxalisib inhibits mTOR as well as another related molecule called PI3K, which has been implicated as contributing to epilepsy. Lee was also part of a team that reported results from a small clinical trial in which epilepsy patients received an mTOR inhibitor called everolimus. The drug didn't outperform a placebo overall, but interestingly, two patients who achieved seizure freedom on everolimus in the last month of the trial carried mTOR mutations, whereas none of the patients with other mutations had this positive outcome. All of this goes to show that studying spontaneous mutations is more than an academic exercise: It's essential to the future of medicine.

■ ■ ■

When people talk about genetic diseases, they are typically referring to inherited ones. Genetics seemed inextricable from heredity for so long.

This outlook traces back to the work of the Austrian friar Gregor Mendel, whose experiments seeded our knowledge about how traits are passed from one generation to the next. He bred peas and tracked the patterns of their shapes and colors, ultimately showing that the plants' attributes were either "dominant" or "recessive." The dominant traits would be passed on to the offspring plants if either parental line possessed them. In contrast, the recessive traits would emerge in the progeny only if *both* parental lines carried that *same* trait. For example, if both parental pea plants had pods that were yellow—a recessive trait—their progeny would, too; otherwise, if even one of the parental plants had green pods, the offspring's would be green as well.

Mendel's work came many decades before the era of modern genetics, but it later became clear that similar rules often apply to the transmission of heritable human diseases. Broadly speaking, we have two sets of chromosomes and therefore two copies of each gene. In some cases, we carry a dominant and recessive version of the same gene in each of our cells. Other times we might have two recessive versions. When this recessive pairing happens, and when it is linked to a disease, we can end up quite sick. A classic example of this is the harrowing lung disease cystic fibrosis. If someone has two mutated copies of the gene called *CFTR* (short for cystic fibrosis transmembrane conductance regulator), their airway cells end up having trouble clearing mucus. That leaves them prone to deadly lung infections. In fact, until the advent of recent treatments in the past decade or so, the life expectancy of someone with cystic fibrosis was around twenty-six years. But if someone has only one mutated *CFTR* gene, they generally don't experience classic symptoms of cystic fibrosis. These individuals are carriers for the disease but often never know it because their second, healthy copy of the gene picks up the slack.

For many years, genetic diseases were viewed in this Mendelian spirit. We talked about them as something passed down from generation to generation within families. *Hereditary* mutations causing long QT syndrome, for example, were described back in the 1990s (this conception of the condition as inherited is what made Astrea's case so mysterious, since neither of her parents were carriers). Indeed, researchers are constantly discovering new heritable diseases. One recent study estimated that around ten thousand such illnesses exist. Even that number, according to some scientists, is likely a "gross underestimate," considering the almost innumerable number of rare diseases that might exist.

Yet—as we've seen with everything from PNH to hemimegalencephaly—we're mistaken if we limit ourselves to thinking of genetic diseases as always inherited.

One genetic ailment that is the result of an event *after* conception—rather than the consequence of a trait passed down from parents—is Sturge–Weber syndrome. A DNA error strikes during early development, resulting in blood vessel abnormalities and a characteristic pink or purple birthmark known as a port-wine stain on the forehead or upper eyelid on one side of the face. Beneath the surface, the genetic mutation causes neurological problems that can affect the eyes and brain. The outcome of the disease can vary widely from patient to patient; some children with the condition experience seizures that worsen over time and developmental delays. All of this happens because of an acquired genetic glitch that occurs from out of the blue while the child is in the womb.

Thanks to modern sequencing, we're learning that many mysterious illnesses are actually noninherited genetic diseases. There's the bone condition melorheostosis, for instance. It was first described in 1922 and causes a painful overgrowth of bone that resembles dripping candle wax on X-rays. Nearly a century after it was first characterized, geneticists uncovered that the mutation behind the disease was surprisingly absent from many cells in patients with the disorder. In some patients, the genetic change was detected only in their bone cells. This nonuniformity overturned the dogma behind the disease. "Scientists previously assumed that the genetic mutations responsible for melorheostosis occurred in all cells of a person with the disorder," one of the doctors behind the study, Timothy Bhattacharyya of the National Institute of Arthritis and Musculoskeletal and Skin Diseases, was quoted as saying about the discovery. The finding emphasized that this was a genetic disease that could arise after conception.

Most cases of Dravet syndrome, a seizure disorder, are due to spontaneous mutations during fetal development in a crucial gene that helps neurons transmit electrical signals. And although we think that acquired mutations in important genes are rare, sometimes lightning strikes twice: Doctors identified a young girl who had *two different* noninherited mutations in the gene linked to Dravet syndrome.

Mainstream scientists are waking up to what a small group of researchers on the cutting edge have been saying for years: that noninherited mutations matter. In 2023, the US National Institutes of Health launched a $140 million effort to try to catalog these previously underappreciated genetic errors in various organs. Some of the first round of grants went to researchers running experiments that included single-cell sequencing of tissues to look for the takeover of mutant clones.

■ ■ ■

Sometimes genetics gets so twisted that you feel you have gone through the looking glass. Consider the fact that there are examples in which noninherited mutations mimic inherited diseases. Take hemophilia, a disorder that makes it hard to form blood clots, which in severe versions can be very deadly. Most cases of hemophilia are inherited, but some are the result of spontaneous genetic changes that affect only a portion of a person's cells. In fact, this latter scenario might happen more often than previously thought.

Then there's a condition called autoimmune lymphoproliferative syndrome, which is known as ALPS. For many years ALPS was assumed to always be an inborn illness. In classical cases of ALPS, the symptoms show up in the first years of life. Unusually high numbers of white blood cells accumulate in places like the lymph nodes and spleen, and there is a risk of anemia. Over time, a patient with ALPS can face problems ranging from hemorrhage to autoimmune complications to cancer. The assumption that ALPS was always inherited was shattered when immunologists uncovered a handful of ALPS patients who had no family history of the disease. In some cases, the mutation was seen in their blood progenitor cells, but not their hair or cheek cells, underscoring that the mutation was not present at their conception. Another study documented more such cases, noting that the noninherited form of ALPS seemed to manifest clinically at an older age than the congenital form.

When a disease is caused by a spontaneous mutation that mimics an inherited one, it is known as a phenocopy. The number of known phenocopies, which do not follow a Mendelian pattern of inheritance, is rapidly increasing. Case in point: The vast majority of phenocopies of inborn immunodeficiencies, such as ALPS, have been identified since the 2010s. (The question facing Astrea's doctors was how they could definitively prove that her heart condition was a phenocopy of long QT.)

Down syndrome, the most common chromosomal disorder diagnosed in the US, is almost always—98 percent of the time—due to an extra copy or duplicated portion of chromosome 21 that accidentally occurs in the egg, or much more rarely the sperm, that forms an individual. However, in many of the remaining cases, the chromosome duplication occurs only *after* the embryo starts growing, so only a fraction of the person's total cells are affected. In other words, in rare cases the extra chromosome is not present at conception. Instead, it crops up when the embryo has already passed into a multicellular stage, meaning only some of the cells that give rise to the body carry the extra copy of the chromosome. Some people with this mosaic form of the disease have all the physical features of Down syndrome but a high IQ. Others with the mosaic form have no outward physical signs but must live with the health and cognitive effects.

Mosaic forms of conditions such as Down syndrome can go undetected for decades. An unexpected case turned up in the analysis of cells from a fifty-five-year-old man who had shown increasing signs of forgetfulness and dementia. He had worked in the navy and as a welder for twenty-five years, and he wasn't having any difficulty in his job as a custodian in a public school. But his wife and family members noticed he was misplacing items and having trouble operating a cell phone and performing many other basic tasks. His physical appearance seemed normal, but doctors noted that his chin was diminutive and his pinky finger on both hands had a bone abnormality, so they ran some genetic tests on his cells. The tests revealed a surprise: The patient's body was a genetic mosaic. Some of his cells had an extra copy of an entire chromosome—chromosome 21. The extra chromosome was present in twenty out of the two hundred cells they examined, indicating that he had a mosaic form of Down syndrome. The discovery was significant because by age forty almost all individuals with Down syndrome have the brain plaques linked to Alzheimer's, and up to half of them subsequently develop an early-onset version of Alzheimer's. The findings explained the man's worsening cognitive issues.

Another stark example of the influence of acquired mutations comes from a medical case report of a woman who began having signs of dementia and mild leg weakness at the age of forty-two. Her symptoms worsened progressively over the course of sixteen years, and she died at fifty-eight years old. It was only after her death and a postmortem autopsy showing telltale plaques in her brain that doctors were able to give a definitive diagnosis of early-onset Alzheimer's. A genetic analysis of her cells revealed a mutation in the gene for presenilin-1, the protein most commonly mutated in inherited forms of Alzheimer's disease that manifest in midlife. What was odd was that the mutation was present in only 14 percent of her sampled brain cells. Through careful calculation, the scientists estimated the genetic error had happened sometime within the first two weeks of her development as an embryo. (That was not the end of her family's tragic story. The patient's daughter began showing different but related symptoms of leg weakness and dementia at age twenty-seven and passed away twelve years after. Researchers believe that this intergenerational recurrence suggests that the mother's spontaneous somatic mutation in presenilin-1 was passed on to the daughter, although they cannot say for certain.)

The list of diseases in which noninherited genetic errors have been implicated keeps growing. These somatic mutations even seem to underlie some cases of autism. Christopher Walsh's group found evidence that as many as one in twenty children with autism spectrum disorder have potentially damaging acquired mutations. Given the rising number of diseases tied to noninherited DNA changes, more and more doctors are saying that medicine should embrace this reality. Recently, Jason Liebowitz, a physician who teaches at Columbia University's medical school, wrote that "if, while seeking diagnoses for patients, we stop and consider the possibility that diseases we already know are linked to somatic mutations, this could help improve our practice."

Even very common diseases for which scientists have long sought inherited risk factors are being reexamined as possibly resulting instead from spontaneous mutations that occur in the body.

A perfect example of this is endometriosis. According to World Health Organization estimates, as many as one in ten women of reproductive age suffer from endometriosis. That ratio translates to 190 million people on the planet right now. Given its sky-high prevalence, the odds are that you or someone you know is affected by the disorder. The condition is characterized by a wide range of symptoms that can include heavy menstruation and debilitating pain in the pelvic area. To add insult to injury, it is also linked to infertility. The disease is known to run in families, but not all cases are hereditary. One study in *The New England Journal of Medicine* described how among two dozen patients with endometriosis, nineteen had cells with acquired mutations in a suite of genes. Of note, many of those affected genes had previously been linked to cancer, but none of these women had that disease. Instead, they had growths resembling the inner lining of the uterus, known as the endometrium, in places far-flung from the womb. The extra tissue had invaded parts of their bodies such as the bladder, appendix, and abdominal wall. Tissue of this sort is not "malignant"—in the way we refer to cancer that can overrun the body and bring about death—but it can cause chronic agony.

The discovery of acquired genetic changes in endometriosis detailed in *The New England Journal of Medicine* inspired other scientists to take a closer look at what was happening in this awful and widespread disease. One group of Japanese researchers decided to pursue the theory that women of reproductive age may be prone to endometriosis-causing mutations because they menstruate. The endometrial lining of the uterus sheds and regenerates each month during a woman's reproductive years. The Japanese scientists wondered if that constant cycle of replenishment set the stage for genetic mistakes to arise and for mutant cellular clones to start outcompeting their neighbors.

The research team went hunting for mutations in the endometrial glands of the uterine lining. They managed to collect genetic sequencing of cells from more than thirteen hundred endometrial glands from several dozen women, including many who had undergone a hysterectomy. The study revealed an overrepresentation of certain mutant clones in a number of the uteruses, suggesting those cells were somehow winning out. For example, in one fifty-year-old woman, two different mutant clones had emerged and occupied large areas of the endometrium. A Darwinian battle seemed to be playing out among her endometrial cells. The study showed the "evolutionary dynamics of mutant clones" in the lining of the uterus, the scientists concluded in their write-up of the experiment. They contended that the overrepresentation of certain mutant cells might be the root cause of endometriosis.

■ ■ ■

People often speak about evolutionary trees that have grand scales stretching across millennia. If you look at the phylogenetics of our own species, you might trace backward from *Homo sapiens* to *Homo erectus*

and *Homo habilis* and, somewhere much farther back, a common ancestor that we share with chimpanzees. We can do a somewhat similar
thing with our family trees. Consider Jennifer Allred, whom we met in
chapter 3. The immune deficiency known as hyper IgM runs in her
family, and she can trace it from her son Adrian through at least six
generations to her great-great-grandmother, who was a carrier of the
disease. Now, with the advent of genetic sequencing approaches that
analyze different cells from the body, researchers are creating evolutionary trees that trace the rise of mutations in a single individual over
a lifetime.

In 2021, a group based out of the Wellcome Sanger Institute announced they had done postmortem genetic analyses of several individuals to reconstruct the genetic histories of their cells. They were
able to mathematically retrace the development of one man who was
seventy-eight years old at the time of his death all the way back to his
first cell division in his mother's womb. The scientists were awed by the
extent of genetic variation in the tissues within each individual.

Some of the fastest—and most high-stakes—cell divisions happen
during the early period of embryonic development. And it's not just the
case for humans. You will find this rapid phase of early growth across
the animal kingdom. In the common fruit fly, the first embryonic divisions of the cell nucleus, where its DNA resides, happen every nine
minutes; by comparison, it takes the average mature cell as long as
twenty-four hours to go through a single cycle. However, rapid growth
during early development comes at a cost: Speedy cell division is linked
to less stability of the genome. It's perhaps no surprise, then, that more
and more research has suggested that a flurry of mutation takes place
right after a baby is conceived.

When a new mutation strikes early in embryonic development, it can be widespread within our tissues, approaching the universal reach of an inherited mutation. If this kind of early, spontaneous mutation disables an important gene, it can have major health consequences.

Thanks to modern genetic sequencing, we're finally beginning to get a grasp of how many somatic mutations occur during fetal development. One study estimated that the stem cells of newborns had experienced almost three times the number of single-letter genetic changes during the nine months in the womb as those of adults had in a year. More recently, an Icelandic study found that identical twins differ on average by around five early-stage genetic changes, further illustrating that the genome is prone to errors from the get-go of embryonic growth. There may be even more genetic changes, such as duplications of DNA sections, that are too tough for current technology to detect.

Beyond the regularly mentioned sources of mutation, such as those arising from damage and poor repair or errors introduced when a cell divides, another source is mobile stretches of DNA. Our cells contain bits of DNA called long interspersed nuclear elements, or LINEs, which can essentially pop out and jump from one genomic region to another—especially during early development.

A particular branch called LINE-1s are thought to be the remnants of ancient virus-like genetic elements that became embedded in the DNA of our ancestral species and have been passed down the mammalian line for at least 160 million years. Some LINE-1s have been active for forty million years in our primate lineage. Often, LINE-1s jump around the genome without consequence. But once every blue moon they cause trouble when they relocate. They might, for example, land in the middle of the sequence of an important gene and disrupt its function.

Rusty Gage, a neuroscientist at the Salk Institute for Biological

Studies on the California coast, and his colleagues have shown that the moving around of LINE-1s results in genetic mosaics. At first, they showed this in the brains of mice. But Gage encountered skepticism that it happened in people. "We knew we were going to run into a sawmill," he recalls. But Gage and his collaborators kept plowing ahead for more evidence. After the rodent study, he and his teammates analyzed postmortem human samples and found hints that LINE-1s were stirring up mutations in the brain tissue of our own species, too. Since then, LINE-1s have been implicated as possibly influencing illnesses ranging from Rett syndrome—a life-shortening neurological condition that causes muscle function deterioration and language loss—to amyotrophic lateral sclerosis, also known as ALS or Lou Gehrig's disease, which causes paralysis and, ultimately, death.

The specialists at Stanford treating Astrea Li knew about the myriad ways that mutations could transpire in cells, and they had circumstantial evidence from her blood samples that her life-threatening cardiac complications were due to long QT syndrome. But they still lacked definitive proof that her heart cells were mutated.

▪ ▪ ▪

While scientists debated about the reach of Astrea's mutation in her body, she was at home, receiving medications and hitting her developmental milestones. She began playing with toys as infants do and soaking in the world around her. But then, one Friday afternoon when she was seven months old, sleeping comfortably in her crib, her mother, Sici Tsoi, got an unexpected phone call. It was the hospital on the line. Although Tsoi hadn't detected anything unusual in Astrea's behavior, the medical team had received a wireless message from the defibrillator in her daughter's heart that it had just shocked the organ back to function. They asked Tsoi to bring Astrea in immediately so they could see what was going on.

Doctors wanted to keep Astrea overnight in the hospital for observation. Tsoi and her husband, Edison Li, were processing everything. "From our eyes she was completely normal. There was nothing out of the ordinary that we could see," Li says. However, the physicians were tracking Astrea's heart function, and it was declining. The organ seemed to be experiencing cardiomyopathy, a condition of heart muscle weakening. They kept her in the hospital. Then, Li recalls, right after midnight on Wednesday, when he and his wife were away from the hospital, they learned that Astrea's heart had stopped. They drove to the cardiac ICU and waited across the room from where around thirty different medical professionals were working to save their daughter's life. "The floor was all blood. That was the hardest three hours we experienced in the whole thing," Li says. Four hours later they could finally see her. She was hooked up to an ECMO machine, which was taking over the functions of her heart and lungs to provide oxygen to her body. A few days later she was disconnected from that, hooked up to an artificial heart, and put on a wait list for a heart transplant.

Luckily, a transplant became available less than two months later. The procedure also offered an opportunity for James Priest, one of the cardiologists, to advance in his quest for answers. He wanted to understand why Astrea had suffered the heart malfunction—the cardiomyopathy—that necessitated her heart transplant. "What was going through my head is, 'What's going on?' There are genetic differences that people can have that cause long QT syndrome and then cause cardiomyopathy. But then there are also times when we give people a pacemaker that can actually induce cardiomyopathy. And so we were trying to figure out which one of those two things might have caused her to have the cardiomyopathy," Priest says. In other words, he needed to know if long QT was truly the basis of Astrea's heart problems. "I talked to the transplant cardiologist and I said, 'Please, we need to get a sample of her heart.'"

Astrea underwent a successful surgery. Even better, she was able to go back home after her hospital recovery and resume a happy childhood with her family. Meanwhile, her heart samples were analyzed, and data was sent across the country to the lab of Natalia Trayanova, a professor of biomedical engineering at Johns Hopkins University in Baltimore. The genetic sequencing had shown that Astrea's heart did have a small population of cells with the *SCN5A* mutation, and Trayanova, along with her graduate student and another colleague, built a computer model of the organ to demonstrate that it was a significant enough portion of cells to cause long QT syndrome.

For Priest, it felt like a huge victory. "Most of the time you have a pretty good idea of what somebody's problem is when they walk in the door or after talking to them for ten minutes," he says. In contrast, "this was a medical mystery, and it turned out to be even more complex and nuanced than we could have imagined." It was the first time that long QT had been shown to be the result of a genetically mosaic heart. Priest describes it as a clarifying moment: "It really led us to this whole new awakening that diseases that we had previously considered to be only Mendelian—meaning inherited from your mom or your dad—or occurring as a mutation in a sperm or an egg, could actually in some cases be caused by mosaicism, which is not something that had been previously recognized, certainly for long QT syndrome but also for a variety of inherited cardiovascular diseases."

Armed with more data, Priest and his collaborators revised their research paper draft describing Astrea's case to include the DNA sequences they obtained from her heart. They also added information about four more people with mosaic, noninherited forms of long QT syndrome that they had subsequently identified. Finally, the paper was ready for publication; it appeared in 2016 in the *Proceedings of the National Academy of Sciences*. It cemented the case that long QT was among the list of diseases that could be caused by a spontaneous, non-inherited mutation.

Astrea's story shows how difficult it can be to demonstrate that mutant cells are causing trouble in tissues that doctors cannot easily access with a biopsy. Despite these challenges, in the years since, there have been other examples of heart disorders caused by spontaneous genetic errors. They include mosaic versions of Marfan syndrome, a disease that causes faulty connective tissue that can weaken the aorta.

Today Astrea is a preteen who is interested in crocheting and making small crafts from YouTube tutorials. Her remarkable story illustrates the triumph of modern medicine over the threat of harmful genetic mutations. Sometimes, though, luck strikes in the absence of treatment options: A patient can overcome an existing life-threatening mutation not with medical therapy but rather through an *additional* chance genetic event.

When Our Bodies Autocorrect

One could argue that Tom Whitham was born to be a plant sleuth. His family had established the very first plant nursery in all of Iowa back in 1863, and they had remained rooted in the business ever since. Whitham, who is in his seventies, speaks with pride about that original nursery site, which is now a conservation park where people can go cross-country skiing or bird-watching among one hundred different species of trees. The family's farming enterprise had relocated by the time he was a kid. His father moved the business to a large area about twenty miles away from the original plot in the 1940s. Whitham spent much of his youth helping his dad grow the bushes and trees that they would sell wholesale. It was a massive operation. The family nurtured some of the trees for several decades, until they grew four stories tall. Whitham recalls how each big tree would need to be transported on one railway car all by itself to allow for its twenty-ton mass and giant root ball. "Banks were the ones that *really* bought a lot of the big trees because they were very expensive," Whitham recalls. "A bank wants to look like it's a very long-lived, ongoing concern. And so they want big trees to make everybody who visits feel like 'Oh, this bank has been here a long time so I can trust it.'"

In the moments Whitham could steal away from helping his father, he would climb up high into the trees on their land and get lost in the science fiction books he loved as a child, by writers like Jack London and Isaac Asimov. His favorite tree to spend hours reading in had a few branches that came together like a chair.

Whitham's duties on the nursery land included looking for branches that were different among the trees, and then working with his father to figure out if the difference was due to the influence of a pest or reflected a change in the weather or soil. If none of those explanations fit the bill, then they knew they had found a unique genetic trait in a plant—one that they could name as a new variety, patent, and potentially sell commercially. To confirm it, they would propagate the plant to see if the trait was stable.

One day in the 1960s, when he was a teen, Whitham was wandering the rows of saplings around his family's hundreds of acres when he stumbled upon an anomalous young tree. He had been strolling through a plot of variegated Norway maples. Normally, these trees have leaves with wide white margins along their edges. But the outlier he spotted had a branch with leaves that were solid green. Whitham didn't know it initially, but the mixed-up tree would grow even stranger with time.

Whitham told his dad about the unusual branch among the variegated Norway maples. His father's first inclination was to prune it away from the tree. The anomalous branch didn't have any particularly striking characteristics besides having leaves that were totally green—which is just like the leaves of most other tree varieties. But out of curiosity the family decided to let it continue to grow unfettered. They wanted to see what would happen.

Month after month, it became clear that the branch with the green buds and leaves was growing faster than those with variegated foliage on the same tree. Compared to their variegated cousins, solid green leaves contain more chlorophyll, the pigment that plants use to convert sunlight, water, and carbon dioxide into food. The greener branch was simply better at feeding itself. This helped explain what was happening with the Whithams' tree. That empowered branch with solid green leaves was staging a takeover of the rest of the plant.

As the years passed, the Whithams watched as the other branches were outcompeted by the green-leafed one, which began growing more vertically toward the sky and sprouting new offshoots. By the end of a decade, 95 percent of the foliage on the tree was green. There were other instances on the farm in which he and his dad spotted green-budded branches among their variegated Norway maples, and they allowed this natural experiment to unfold again. Each time, the green branch on a tree would win out.

Whitham went on to get an undergraduate degree in horticulture and ultimately became a university professor in biology. He has authored more than 270 scientific papers, which have appeared in prestigious research journals like *Nature* and *Science*. One of his publications was a 1981 report in which he and a university colleague reflected on what Whitham and his father had learned from forty years of observing anomalous branches on their Iowa nursery. "The smaller the tree, the faster the changeover" from variegated to green, once a rogue branch turns up, they wrote.

Weird vegetative takeovers don't just crop up on farms; they happen in home yards as well. Sometimes, the case involves a dwarf Alberta spruce. This kind of tree should normally grow slowly and stay compact, but once in a while, seemingly out of nowhere, a section of it will flourish faster than the rest. According to Bert Cregg, a professor of horticulture at Michigan State University, the complaint is usually something along the lines of "There's a tree growing out of my tree!"

The strange events in which anomalous branches take over are often seen in special plants, such as those with variegated leaves or dwarfed traits. These ornamental cultivars are the result of a genetic mutation to begin with, and a second, reverse mutation in part of these plants can cause that portion to revert back to normal growth.

Part of the appeal of studying revertant mutations in trees is that unlike in the human body, where cells grow in squishy tissues or circulate in blood, plants have more static tissue. You can find new genetic changes in one part of the plant and easily follow them as they are passed on to the subsequent branches that sprout off from that point. For example, a group of Swiss scientists at the University of Lausanne examined the famous 234-year-old "Napoleon" oak tree on their campus—so named because it was already growing there when the French general and his troops crossed the grounds—and were able to trace the mutations that were passed from older parts of the plant to its new, younger branches.

Trees are a magnificent example of how a single organism can accumulate mutations over time. The sheer genetic diversity in a single long-lived tree might surprise you. A group in Canada did a genetic analysis of bark and needle samples from twenty Sitka spruce trees growing on Vancouver Island and estimated that a single tree could have up to one hundred thousand genetic differences between its base and the tip of its crown. Meanwhile, a recent study from another group suggested that, among trees of the same size, older ones have accumulated more mutations per meter than their younger but equally big counterparts. Because of this, the researchers concluded that long-lived trees might possess more genetic adaptations to their environment that boost the chances of their species' survival.

These findings support a theory that Tom Whitham wrote about in the early 1980s. He contributed to what became known as the "genetic mosaic theory" of plant defense. The idea was basically that trees have the opportunity over many years to collect mutations that protect their newer parts against pests. Whitham gave the example of narrowleaf cottonwood trees, in which acquired genetic variations seemed to confer certain branches safety against invading aphids.

Less than a decade after Whitham published his ideas about plant defense, Australian scientists discovered a striking case of a beneficial mutation rescuing a tree. The researchers were exploring a section of New South Wales that had experienced a beetle outbreak when they happened upon a strange-looking eucalyptus tree. The tree had been completely decimated by pests, which had eaten all the leaves—save for one section at the top. The immune leaves belonged to a lone branch that reached for the sky. It remained virtually untouched, whereas the other branches were infested with beetles. In the early 2010s, another group of scientists analyzed samples from the tree and found ten specific DNA changes that were present in cells of that special branch but not the rest of the tree. This gave hard evidence that genetic changes were responsible for the branch's resilience.

Mutations can have miraculous effects. They can help an organism thrive beyond its initial genetic destiny. But they don't just rescue plants. If you look close enough, you can find examples of the phenomenon of revertant mutation happening in other species, including humans.

■ ■ ■

Around the time that plant scientists were marveling at the resurrected branches of trees, medical researchers began finding glimmers of corrective mutations in human tissues. One geneticist who did so is Eric Hoffman. In 1986, when he was in his late twenties, he joined the team that shortly thereafter characterized the complete gene affected in Duchenne muscular dystrophy, a hereditary degenerative disease. The illness typically causes death in early adulthood from complications affecting breathing or the heart. The gene that was identified as the culprit normally produces a key protein—which was dubbed dystrophin—that helps stabilize cells. In Duchenne, the gene is broken, and muscle fibers that lack the stabilizing protein degrade over time from the wear and tear of contracting.

To better understand the illness, Hoffman turned to a strain of mice with a version of Duchenne. He spent hours closely studying the animals' biopsied muscle tissue under the microscope in the lab. Using a special stain that emits a yellow light when it binds to dystrophin, he sought to confirm that the rodents' muscle fibers didn't produce the protein. Normally, with Duchenne muscular dystrophy, there aren't any cells that pick up the dye because the body cannot make dystrophin. So Hoffman expected the slides to be completely black. But when he peered through the microscope, he was stunned at what he saw. Quite a few cells were lighting up yellow. Hoffman was based at the Boston Children's Hospital, and he phoned his collaborator Simon Watkins, who was working across the street at the Dana-Farber Cancer Institute. "I called him over and said, 'Simon, what's this?'" he recalls. His colleague was unfazed by the dystrophin-making cells. "He came over and looked and said, 'Oh, I see those all the time.' And I said, '*You see these all the time?*'"

Hoffman questioned his colleague further, hoping for an explanation about what the fibers represented. Watkins didn't have an answer. He didn't know. Yet the presence of the cells continued to fascinate

Hoffman. Interestingly, exposing the mice's legs to radiation—a known mutagen—increased the number of cells with restored dystrophin, suggesting that a genetic change was behind the recovery. Together with Watkins and a team of fellow researchers, Hoffman published the findings in 1990. "That paper was one of the fastest I ever wrote," Hoffman says. "I was going between talks in Germany, and I distinctly remember I had a couple of hours on a train and I wrote it from beginning to end."

Before the mouse study, Hoffman and his collaborators, as well as other research teams, had occasionally noticed a faint signal of dystrophin production in muscle tissue from people with Duchenne muscular dystrophy. But it seemed minimal. In the decades that followed, scientists searched more closely for self-healed cells in their patients with this condition. One group showed that people with the disorder would experience an increase in the number of corrected fibers up until their early teens. They found clusters of as many as twenty corrected fibers in the patients of older age. This suggested that a genetic fix enabling a cell to make dystrophin could be passed on to its daughter cells.

There were even cases in which self-corrected cells appeared early enough in a person's development to make a meaningful difference to the clinical course of their disease. For example, Hoffman and his colleagues reported the unusual case of a young man with Duchenne. The patient had three uncles with the condition who had all died in early adulthood. He also had the expected progressive weakness associated with the disease—but strangely, it mostly affected only his left side. A close-up of his DNA revealed that about a quarter of the cells on his right side had overcome the offending genetic error. Given how many reverted cells he had, it was most likely that this correction had happened in a cell very early in his development, at the embryonic stage.

Knowing that the body can mutate to repair or rescue its tissues is useful. Although it's exceptional for self-corrected cells to change the clinical course of disease for most patients with Duchenne muscular dystrophy, DNA changes that resurrect gene function in rare diseases merit study because they might inspire new treatments.

The prospect of self-corrected cells saving the day is greater for some conditions than others. For illnesses affecting muscle, it's a tall order after the body develops, because there's less cell turnover in muscle compared with other elements of the body, such as blood. That's particularly true in adulthood: "In Duchenne, everything's locked in place in the tissue," Hoffman explains. "If you have 99.9 percent of cells [with the disease mutation] and you have one individual cell [that has reverted to normal], that's not relevant to the patient." In patients who have passed adolescence, this makes it difficult for a self-healed cell to reach meaningful numbers such that it can rescue muscle function. Instead, it remains stuck, holding the secret to a cure but unable to help. But after this discovery in muscular dystrophy, examples of self-correcting cells in other diseases kept coming. And the message of these observations was increasingly a hopeful one.

■ ■ ■

In the early 1990s, a clinical geneticist named Eli Anne Kvittingen was examining the liver cells of her young patients in Oslo when she stumbled upon a strange anomaly. At the time, she had been studying a hereditary form of a disease called tyrosinemia for around a decade. The severe form of tyrosinemia she was investigating involves a genetic defect that makes it difficult for the body to process certain proteins in food. The result is a buildup of a toxic metabolite that progressively destroys the kidneys and liver. Infants with the acute form of the disease who fail to receive treatment die within the first year of life from organ failure. There were very few therapies available when Kvittingen was conducting her research, so some young children with the disease would receive liver transplants in order to survive.

One side product of the transplant procedure was that researchers like Kvittingen could examine the diseased livers that were removed from the tyrosinemia patients to make room for the healthy donor organs. She ran tests on the discarded livers from a handful of such patients between one and six years of age and noticed something peculiar. Amid the swaths of malfunctioning liver cells, there were clusters of cells that possessed the enzymes necessary to break down the toxic metabolite. Kvittingen and her colleagues published the stunning result in 1993, concluding that these unexpected islands of cells "reflect a correction of the enzyme defect in clones" of replicating liver cells that had somehow overcome the deadly tyrosinemia mutation. It was, as the title of their paper proclaimed, a "self-induced correction."

We hear so often of rare, intractable genetic diseases that it is jarring to think that the body could happen to overcome them with an autocorrection. Yet that was precisely the hopeful message that Kvittingen's study offered. Even though the small islands of healthy cells in the children's livers had not grown quickly enough in number to save the young patients from the need for transplantation, their mere existence was tantalizing. The upshot was that some people born with a fatal genetic disease had cells inside them that had somehow become perfectly normal.

What Kvittingen and her teammates did next was key. In a follow-up study published a year later, they pinpointed the precise DNA changes that made the spontaneous recovery of the cells they found in the diseased livers of tyrosinemia patients possible. They showed that those healthy cells had somehow fixed a tiny typo in the DNA sequence that can cause the disorder. Thanks to that fix, the cells could churn out working copies of the enzyme needed to process protein properly—and thereby could avoid a buildup of toxins. Put another way, the cells with the tyrosinemia mutation had found a way to mutate again, *back to normal*.

A trend began to emerge. More and more unusual cases involving other disorders hinted that a body with damaging mutations could sometimes self-correct.

Scientists in Texas reported the case of a man whose "level of intelligence [was] within the normal range" despite the fact that he had Lesch–Nyhan syndrome, a genetic condition that causes severe intellectual disability. Also, even though the disease normally triggers self-harming behaviors such as head banging, the man lacked any compulsion to injure himself (he did, however, have other telltale symptoms, such as kidney stones and jerky movements). When researcher Tom Caskey and his team in Texas studied the man's blood cells in a laboratory dish, they had a high propensity to revert the gene linked to Lesch–Nyhan back to normal. "We suggest that the ability to spontaneously revert this mutation may be responsible for the unusually mild symptoms observed in this patient," the scientists wrote in the paper describing the unusual case.

Then there was the mystery of the two boys who were being followed in New York for a rare immunological disorder. The technical name of the condition they had is adenosine deaminase-deficient severe combined immunodeficiency (ADA-SCID). However, most people know it as "bubble boy disease." It gets this name because those who are born with it have such weak immune systems that they sometimes have to be sequestered within a protective barrier shielding them from any germs.

Both of the boys in New York faced awful circumstances. One of them lacked a matched sibling donor and therefore could not receive a bone marrow transplant, which was at the time the only known effective therapy. The other patient—whose brother had died from ADA-SCID as a toddler—did not receive any therapy for the disease because his parents objected for religious reasons. But then the unexpected happened.

Instead of dying in early childhood as patients with the disease typically do, the two boys got better with each passing year.

Rochelle Hirschhorn, a clinical geneticist in New York, set out with her colleagues to do an intensive analysis to understand this seemingly miraculous outcome. She was able to obtain cells from the first of the two boys at different stages of his childhood and teenage years. She found that while none of his cells at age two produced the vital protein called adenosine deaminase, half of his cells obtained at age sixteen did so. He was still free of any medical problems when doctors reached out to him at age twenty. She and her colleagues reported the findings in 1994, but it was still a mystery how the other patient had survived.

Finally, in 1996, Hirschhorn's group reported that they had cracked the case: They obtained blood samples from the second boy, and from his parents for comparison, and showed that the mutated gene had repaired itself in many of his cells. In an article published alongside the study, the geneticist Hagop Youssoufian wrote that Hirschhorn and her coworkers had provided "a provocative molecular tale of a patient . . . who apparently transcended the bounds of his inheritance."

If a second mutation in the body can override the illness caused by an initial one, how does it end up present in enough cells to make a clinical difference? The prospect of knowing the answer to this question was motivating. If scientists could figure out how the body healed itself, perhaps it could inspire new therapies for tough-to-treat rare conditions.

Hirschhorn and Youssoufian, along with others, began to suspect there were some sort of Darwinian forces at play. The cells that had reverted to normal in the young patient born with ADA-SCID seemed to be replacing the infirm cells circulating in his blood.

The idea that self-healed cells in patients with genetic disorders might have an existential advantage proved prescient.

In the late 1990s, an international team of researchers reported the curious case of a twenty-eight-year-old woman with a painful inherited skin disorder who seemed to be miraculously healing. A brother born before her had died at age one of complications of the disease, called epidermolysis bullosa, which causes severe blistering that can lead to dangerous infections. The hand of the young woman was covered in red sores and inflamed, peeling skin, as expected from the condition, but there were also several large areas that were smooth and lacking any signs of irritation.

When the scientists gently rubbed her afflicted skin, it would blister. In contrast, the healthy patches were completely resistant to any such disturbance. She had had many of the smooth, normal patches as long as she could remember, and many of them had stayed the same size, but other healthy patches were spreading over the years. The researchers studied her cells by culturing them in a lab dish and saw hints that her healthy cells had some sort of Darwinian "selective growth advantage."

Two years later, another team found a fifty-six-year-old woman from Austria with the same disease who also had mysterious healthy patches of skin. Notably, the reversions seemed to occur on her hands and arms, areas of the skin that are often exposed to the sun. Some experts in the field theorized that ultraviolet radiation from the sun had been the key to fixing her cells. The sun's rays are well known to cause DNA mutations that kick-start skin cancer. But in this patient, the sun might have triggered DNA changes that rescued her cells. Other theories also emerged about how revertant mutations occur. Some researchers noted that they might happen in diseases where DNA is more prone to mutate overall, or where the affected gene has a "mutational hotspot."

Scientists reporting on the cases of self-healed patients began speaking of them as examples of "natural gene therapy." They felt hopeful. The data suggested that a small number of reverted, healthy cells could sometimes seed a meaningful recovery. The two boys with ADA-SCID who had defied their diagnosis and continued to get better were examples of this. The children each appeared to have had one initial cell that fixed itself, and that was enough to seed a recovery. A single cell could set a rescue operation into motion by replicating and replacing malfunctioning cells.

Unlike solid tissues, in which self-corrected cells are not as free to multiply and spread to save the day, in inborn diseases affecting tissues with a lot of cell turnover, there is plenty of leeway for spontaneously corrected cells to stage a curative takeover. This is true for genetic disorders affecting immune cells and even skin cells. It's also particularly true for blood disorders. The constant flow and churn of blood cells offers opportunities for new-and-improved cells to replicate and rise to dominance in a meaningful way.

One of the most stunning examples of how a single or a few corrected cells might cause a lifesaving turnaround happens in an illness called Fanconi anemia. This hereditary disease typically causes the body's bone marrow to begin failing, eventually leading to a shortage of blood cells. Some affected people in developed countries have access to marrow transplants that can extend their survival, but if left untreated the disorder can cause death before age ten. As many as 20 percent of individuals with Fanconi anemia show signs of self-corrected cells. However, only around 5 percent have the reversions in their stem cells, which is necessary for a sustained recovery from the genetic disease. When this does occur, they can have complete remission. In fact, the disorder may no longer be detectable in them when doctors test their blood. A German team studying these patients wrote that a "selective advantage of bone marrow precursor cells" carrying the restored gene might explain how patients overcame the condition. In other words, the stem cells that regained a working version of the gene were winning out against the others in a Darwinian race.

But what about health problems that are not rare or inherited? It turns out that cells with beneficial mutations can appear then, too. When a previously healthy body starts to decline, sometimes cells evolve to cope.

■ ■ ■

The liver continues to be one of the best teachers of what the body is capable of. If you lost 70 percent of your liver, the remaining 30 percent would grow back to almost the original organ size within a year and continue plugging away at doing its vital tasks: detoxifying the blood and helping to maintain healthy blood sugar levels, among others. The liver is the sole solid internal organ you possess that is capable of full regeneration from a small piece back to its full size.

Consider what happens when a living person donates part of their liver. Doctors might remove around 60 percent of the liver, and the organ will essentially regenerate back to its previous size within the subsequent year. But don't be fooled into thinking that the organ regenerates only when a major chunk is lopped off. Regeneration in the liver is happening all the time to everyone, even those who don't undergo big surgeries. "You have little bits of your liver dying off all the time," explains Hao Zhu, a physician at the University of Texas Southwestern Medical Center who has devoted most of his career to studying the organ. The organ needs to replace cells as part of its regular maintenance or sometimes as the result of minor insults, such as cell damage and loss after someone consumes too much alcohol at a Christmas party or even from excessive use of certain herbal supplements. "The liver has evolved to be a regenerator because humans are accustomed to injuring it," Zhu says. "It's on the front lines of dealing with all the poisons that we ingest."

The liver's capacity for regeneration has been appreciated from ancient times to the present. It was even immortalized in Greek mythology in the story of Prometheus. Among the titans, he was a particularly sympathetic figure. He had a soft spot for helping humans, and it got him into trouble with Zeus. As the story goes, Prometheus raided the workshop on Mount Olympus to steal fire and give it to humanity so they could cook and keep warm. Zeus was furious and handed down a cruel punishment. Prometheus was chained to a cliff in the mountains of the Caucasus where for thousands of years an eagle would arrive daily and feast on his liver. Each night the organ would regrow, ready to be devoured by his winged tormentor once again.

A 1931 rat experiment proved that the liver's ability to bounce back is not mere myth. George Higgins, a scientist at the Mayo Foundation, along with the surgical fellow Reuben Anderson, removed two-thirds of each rodent's liver and observed that the organ returned to its normal size within around ten days. It was a stunning recovery. Doctors had seen patients bounce back from surgical or chemical injury to the liver in the past, but Higgins and Anderson's rat experiment finally gave precise evidence of the extent to which the organ could restore itself.

As the decades have passed, the regenerative power of the liver has remained no less fascinating. Zhu wanted to look for islands of resilient cells in the liver that help to repopulate the organ. Perhaps, he and his collaborators thought, those cells produce more copies—or clones—of themselves than their neighbors because they have evolved some sort of genetic advantage. They reasoned that all the toxins the liver encounters might create a tough environment that heightens the advantages of the resilient cells and their copies. "The liver is one of those tissues that is being chronically damaged and regenerating," Zhu explains. This made him and his colleagues believe it was a prime environment for populations of mutant cells to grow in numbers: "We felt like the clones would get bigger in a faster way."

One of the places Zhu and his team decided to look was in the livers of people with cirrhosis. The liver is usually a smooth organ about as big as a medium-size papaya and sits in the upper right section of your abdomen, but in cirrhosis the organ has a scarred, bumpy surface more akin to an American football. The scarring in cirrhosis is a by-product of the liver's attempts to heal itself, which are not always successful. Drinking too much alcohol can cause cirrhosis over time, as can a wide range of other culprits, including obesity and viruses such as hepatitis B and C. The condition is more common than you might think: It's associated with more than one out of every fifty deaths worldwide.

Zhu and his collaborators went searching for mutant cells with regenerative powers in cirrhotic livers. They analyzed the genetic makeup of cells from the livers of seventy-five people with the condition, and compared this against the DNA sequences of the participants' blood cells, which acted as a reference. They also examined the mutations in cells from six cirrhotic livers that had been totally removed using transplant procedures (as well as one completely normal liver for comparison). The scientists were encouraged to find recurrent mutations in the cirrhotic livers—and cells with these genetic changes seemed to repopulate the damaged liver better than their surrounding counterparts. They gave more evidence to support this by mimicking these mutations in mice and showing that those rodents were better protected from liver damage. Zhu's team published the results in 2019. He had found what he was hoping for—proof that the body mutates while regenerating its damaged tissues.

Traditionally, genetic mutations in the liver that increased cells' fitness had been considered to be a hallmark of cancer, but these findings helped shift that view. "It now appears that not all of these mutations are harbingers of future cancer," the scientists Miryam Müller, Stuart Forbes, and Thomas Bird wrote in a piece commenting on the study Zhu helped lead. "By contrast, some may in fact aid regeneration and protect from further injury."

Zhu's group had shown that cirrhotic livers seemed to be fertile ground for cells with genetic changes linked to regeneration. As fascinating as this was, it didn't prove that those DNA changes were actually working to reverse the disease.

In 2021, not long after his team's study came out, Zhu happened to read a paper from another group that piqued his curiosity. Across the ocean, a group of geneticists led by Peter Campbell, then of the Wellcome Sanger Institute, had zeroed in on intriguing mutations that crop up in fatty liver disease, a condition linked to high weight gain and marked by a buildup of fat in the organ. The disease is on a lot of doctors' radars. It has increased by more than 50 percent globally and affects about a third of the population worldwide, including, by some estimates, around one hundred million people in the United States.

Campbell and his colleagues had analyzed liver tissue from thirty-four people, many of whom had fatty liver disease. The data strongly suggested that cells with metabolically relevant mutations were winning out in these individuals. Some of the mutations were in genes involved in handling insulin, an important messenger in how the body stores and leverages the calories we consume. Other mutations were repeatedly seen in genes that help cells hang on to molecules of fat.

What made the results even more intriguing is that the scientists kept coming across the same kinds of gene mutations again and again across the group. In the language of evolutionary science, they were finding *convergent* mutations—a term that describes when different genetic solutions to the same problem arise totally independently in disparate organisms or distinct cells. When this happens in nature, it points to powerful environmental factors influencing the evolution of unrelated species. When it happens in a human body—such as in individuals with fatty liver disease—it suggests that there was some strong selective pressure going on for those particular mutations to arise.

Back in Texas, Zhu read the results from Campbell's team and his mind started racing: It had previously been clear that evolutionary forces were causing cells with certain mutations to flourish in a compromised liver, but now it seemed that they might also *protect* against the causes of liver disease. Zhu reached out to Campbell. Maybe they could work together to find out if the liver's mutant clones really can fight back against disease? Campbell agreed. All they had to do next was design the right experiment to find out.

▪ ▪ ▪

Zhu and Campbell decided that the best way to test the might of mutant liver cells was using bioengineered mice. The first step involved creating a group of mice, each of which had a subset of liver cells with mutations in sixty-three genes that mirrored genetic changes seen in fatty liver disease. The results echoed what they had suspected happens in humans: Cells with these genetic changes became more prevalent in the rodents' livers when they were fed chow that mimicked a high-calorie Western diet. This outcome suggested that—when faced with an onslaught of overnutrition and a bombardment of fat—the liver experiences genetic alterations that help it cope. "In fatty liver disease, there is a specificity to the mutations where they are directly dealing with the disease that the person has," Zhu told me.

The scientists then did a more exploratory test. They ran the same protocol in mice but this time changed a broader range of genes—472, to be exact—many of which had never been implicated in fatty liver disease. (Usually changes that disrupt the function of a gene can be bad for health. But in some cases they can actually help.) Tweaks to a few of these genes, including one called *Smyd2*, seemed to give cells an advantage in a high-fat environment. Perhaps, they thought, these could be additional healing genetic changes in fatty liver disease?

Zhu and Campbell knew that even if liver cells could independently evolve a way to cope with excess fat in the mice, there was a sinister culprit: time. The cells simply couldn't replicate at the necessary speed to overtake the organ and save it. (A similar problem is thought to occur in humans.)

To see whether the beneficial effect of a mutation would be more pronounced if it was widespread in the liver, rather than just in tiny islands of cells, Zhu and his collaborators engineered mice that lacked the *Smyd2* gene throughout the *entire* organ. Doing so resulted in a clinically meaningful difference. The rodents' livers accumulated less fat than those of genetically unmodified mice and showed fewer signs of tissue damage. It protected them from developing fatty liver disease.

The next question was obvious: What if there were a way to imitate this organ-wide genetic change in humans? The researchers searched through previously published research and were thrilled to find that an experimental compound that could inhibit the *SMYD2* gene (the human version of *Smyd2*) already existed. It was called AZ505. With this compound in hand, they could mimic the protective mutation that silenced the gene across the entire liver in each animal using a drug, rather than through elaborate bioengineering. It worked like a charm: When they gave mice AZ505, the compound shielded the rodents from fatty liver disease. This molecule was imitating the healing genetic change, and it was working on a scale big enough to actually stop the illness. Zhu dreams that drugs found in this approach will work in humans, too. It's no slam dunk yet, though: There are many hurdles, including showing that such drugs are safe and effective in people.

Ultimately, Zhu's liver research has made him see mutations in a more positive light. He says that the beneficial mutations that crop up in people with fatty liver disease show that these genetic changes are not always a bad thing. "I think it's exciting because it gives you a sense of optimism and hope that our bodies are trying to figure out ways to deal with disease," he says. "It's not always successful, but there's a glimmer of hope there where a small part of our body *is* successful. And then we can learn from it." In other words, small populations of cells that evolve resistance to a disease might not always be able to save the whole body. But their genetic uniqueness can inspire drugs that are capable of this feat. Zhu and Campbell are among the founders of a company, Quotient Therapeutics, that hopes to find drugs inspired by noninherited mutations, those that arise in the body. In late 2023, they announced that they had successfully raised $50 million in funding.

Meanwhile, other scientists are trying a totally different tack to treat liver disease. One that's more radical.

■ ■ ■

For patients experiencing dire liver failure, a transplant is the best option. But there are big challenges associated with this treatment. To start with, there is a chronic shortage of livers for the procedure. And people who receive such transplants have to take tricky immunosuppressant drugs—which reduce the chances that their body will reject the donor organ but can raise the risk for infection—for the remainder of their lives. That's why liver transplants are typically limited to the most acute patients. For these reasons, scientists have explored the idea of treating patients with liver disease using small batches of healthy cells that can replenish the organ from within. If it works, it could provide a way to improve liver function for a wider range of illnesses involving the organ, including cases where patients are not sick enough to warrant the risk of undergoing a major surgery. It could perhaps also cut the need for drugs to prevent organ rejection.

There's a major hurdle to overcome, though: Despite the fact that the liver can regenerate, sometimes healthy cells can't get a leg up over their struggling counterparts in the organ. Recall the young tyrosinemia patients Eli Anne Kvittingen identified, whose islands of corrected cells in their livers didn't grow fast enough to save them from needing a transplant. What if there was a way to make it easier for helpful cells to expand more prolifically within solid organs? It's no longer a theoretical question. Researchers are trying this precise approach in experimental settings. They are designing treatments that give cells with healing powers a growth advantage so that these cells *can* rise up in sufficient numbers to stop disease.

One person leading the charge in this radical approach is Markus Grompe, director of the Oregon Stem Cell Center at the state's health and research university. Back in the 1990s, Grompe conducted animal and human studies that shed light on mutations in the gene linked to a severe form of tyrosinemia, the same inherited condition Eli Anne Kvittingen was investigating in her young patients. Grompe recalls reading Kvittingen's 1993 paper about the rare liver cells that regained normal function in a handful of tyrosinemia patients. He was impressed by the finding. It fueled his ambition to find a new way to help those born with the disease.

Just a few years later, Grompe and his colleagues published a paper with a remarkable result. They had been studying mice that were born with the tyrosinemia gene defect in all their cells. These mice would die without treatment. But the scientists transplanted ten thousand normal liver cells into nine of the mice. All the mice that received these healthy liver cells survived. Even more surprisingly, when the scientists transplanted as few as one thousand such cells into six mice, five of these rodents survived. A thousand cells were also enough to repopulate the mutant rodents' livers with functioning tissue.

The results, according to Grompe and his coauthors, demonstrated the "strong competitive growth advantage" of the transplanted normal cells over the animals' mutant cells. Further tests suggested that after two months, the proportion of healthy cells in the animals' livers had reached 90 percent. Grompe subsequently called this kind of treatment "therapeutic liver repopulation."

Grompe and his collaborators were not the only ones curious about transplanted liver cells. A decade later, scientists in New York published data showing how aggressive some transplanted liver cells can be. When fetal stem cells from rats were transplanted into adult rats, they replaced the adult liver cells by actively inducing their cell death.

Liver cell transplantation has been tried in more than one hundred human patients to date. It has helped to partially correct some recipients' liver disorders, but so far the procedure hasn't resulted in a successful long-term reversal of their conditions. This is in part because it's often difficult for these small batches of cells to replace enough of the recipient's liver mass. But Grompe sees a potential solution to this problem. We simply need to give the transplanted cells a better way to take over.

To prove this concept, Grompe zeroed in on a particularly tricky disease: phenylketonuria, or PKU. While rare overall, it's the most commonly inherited disorder involving an inability to properly metabolize a compound found in proteins. Over time, a toxic buildup of the molecule phenylalanine causes damage to the nervous system, leading to permanent intellectual disability as well as other possible complications such as seizures. A severely limited diet, supplemented with specialty medical shakes, can avoid these outcomes, but it is expensive and cumbersome to maintain. Even a simple sandwich can present danger to someone with PKU.

PKU patients lack an enzyme normally found in liver cells that helps break down phenylalanine. Liver cell transplants with a working version of the enzyme have been tried in the past as a fix for these patients, but they seem to help only a bit.

Grompe knew that the dynamics of liver cell transplantation wasn't the same for every metabolic disease. He had shown early on that healthy liver cells transplanted into mice with a rodent version of the disease tyrosinemia would readily outgrow the animals' defective liver cells. But that was largely possible because cells with the tyrosinemia defect are prone to damage, and so their replication lags. In contrast, the genetic defect in PKU doesn't slow liver cells down.

To treat PKU, Grompe needed to give the transplanted healthy cells a leg up. Grompe and his teammates found inspiration for their solution in a compound found in almost every family's medicine cabinet. It's likely that right now you have a bottle of it in your home. To pharmacists, it's known as acetaminophen. But in countries such as the United States and Canada, you're more likely to know it by its brand name: Tylenol. What you might not know is that although acetaminophen does a great job of relieving pain and reducing fever, it can also cause damage to liver cells if taken at high doses for too long. Grompe's team saw a silver lining in this, though. They decided to design healthy liver cells for transplant that were impervious to the damage from acetaminophen. That would be the built-in advantage the cells possessed— and the scientists conducted an animal experiment to prove it could work.

The researchers took healthy liver cells from normal mice and engineered those cells to lack an enzyme called Cypor, which is involved in turning acetaminophen into its toxic by-products. Without Cypor, the cells didn't make the toxic compound. Next, they transplanted these Cypor-lacking healthy cells into mice with a version of PKU. For at least fifty days, the mice received chow that contained acetaminophen. Over the course of the experiment, the proportion of transplanted healthy cells in the animals' livers went from less than 1 percent to 14 percent on average. Grompe and his colleagues saw a complete correction of the phenylalanine levels at a threshold of 8 percent replacement. (They believe that the percentage of Cypor-lacking cells in the liver is low enough that acetaminophen would still work in its regular therapeutic role.)

The mice that received the engineered liver cells had blood test results suggesting that the transplant had helped them overcome their PKU. And not only did the bloodwork show that their PKU had been corrected; their coat color darkened from the flax-brown hue of PKU mice to black, which was the coat color of the healthy control mice in the experiment. It was visual confirmation that the treatment worked.

In their paper describing the results, Grompe and his coauthors refer to the transplantation of the engineered liver cells with the acetaminophen treatment as "selection-enhanced." It's a nod to how scientists recognize that the forces of selection—put forth by Darwin more than 150 years ago—can be inspiration for modern-day medical treatments. Grompe sees a survival of the fittest playing out in the liver, and tries to get his team to understand this. "I tell people who work with me that the human body or any large animal organism is an evolutionary system," he told me.

Ultimately, the goal of these new therapies is to tip the balance in favor of healthy cells in the body.

Sometimes the healthy cells used in therapies arise in patients themselves. There are, after all, rare instances in which a patient's *own* body comes up with self-corrected "revertant" cells that can cure their illness. Doctors have attempted to help these cells grow.

Recall the case of the twenty-eight-year-old woman with epidermolysis bullosa mentioned earlier in this chapter. Her inherited disorder causes severe skin blistering that can lead to dangerous infections. She was the first person with this disorder in whom revertant cells were discovered.

Later, she also became the first person to receive revertant cell therapy in a clinical setting.

Researchers took samples of her cells that had self-corrected and grew them into larger sheets of skin, each about the size of a Post-it note. Scientists removed a thin upper layer of skin about that same size from her thigh and then set the graft in its place. Several months later, the graft had healed beautifully. But less than 3 percent of the cells within it remained corrected. The grafted skin still blistered. It was a sign that the revertant therapy approach needed to be optimized in order to work in patients. There's still a long way to go.

Despite the challenges of treating diseases with transplants of healthy cells, scientists remain optimistic. Recent decades have taught us how unexpectedly resilient the body can be, and how cells can evolve inside us to try to mend what ails us. In certain cases, such as with Fanconi anemia, a few revertant cells can totally overcome a harrowing genetic condition. Other times, like in fatty liver disease, cells with DNA changes emerge to cope with an acquired illness. These DIY solutions from the body are inspiration for new medicines that mimic beneficial mutations, like the drugs that Hao Zhu and Peter Campbell are developing.

All the examples of mutant cells pushing back against disease should offer us hope and reassurance. We so often think of mutation in a negative light, but sometimes it can be a force for healing. Sometimes, mutation is what saves us, and the solution to our health woes comes from within.

7.

The Next Generation

L ike astronomers who use telescopes to peer into the sky for clues
about the dawn of the universe, and geologists who dig deep into
the ground for hints about how the Earth formed, geneticists are
not immune to the pull of origin stories. They pore over sequence data
from present-day humans and re-create theoretical lineages that posit
when different ailments crept into our species. Through these methods, some have calculated, for example, that fifty-two thousand years
ago or so, midway through the Stone Age in Europe, one of our human
ancestors was born with a mutation that causes cystic fibrosis. Nowa-
days, it's the most common life-threatening inherited genetic disease in
the United States. Other teams have suggested that sickle cell disease,
one of the most frequent heritable blood disorders in the world, dates
back to between twenty-two thousand and seven thousand years ago—
perhaps to an individual living in northern Africa or the rainforests of
what is now known as Cameroon. Meanwhile, some researchers reckon
that the most recent common ancestor carrying a prevalent mutation
for Tay–Sachs disease—which progressively destroys the nerves of the
brain and spinal cord—lived around twelve hundred years ago.

Scientists have generally viewed mutation in reproductive cells—which transmit traits to future generations—as something that happens, but happens seldom. They have thought broadly about the chance advantageous DNA changes that occasionally help spur the evolution of new species, for example. And likewise, the emergence of disadvantageous mutations for devastating conditions such as cystic fibrosis, sickle cell, and Tay–Sachs seemed for so long to be like huge but infrequent earthquakes in the genome. In other words, they seemed rare. To this day, many researchers still view reproductive cells—which form the "germ line" that passes genetic material from generation to generation—as relatively impervious to new DNA changes. "I think that this is the stereotype that many people have," explains Kateryna Makova, a scientist at Penn State University. "They think that the germ line is very fixed."

And, in fact, even when new mutations *do* occur in our reproductive sperm and egg cells, the vast majority likely have zero impact. They usually land in the extensive unimportant tracts of the genome. "One important thing to realize is that most mutations are just neutral," Makova says. This understanding was introduced in the late 1960s by the Japanese biologist Motoo Kimura. He argued that a lot of evolutionary change is simply the result of random genetic drift, and that many of the tiny changes in the genome don't affect what our DNA actually produces.

But consequential DNA errors do sometimes crop up in an individual's reproductive cells during their lifetime. And in recent years, many scientists have gained a greater appreciation of the importance of this phenomenon. Much like those astronomers who now, equipped with high-tech telescopes, can spot the birth of numerous new stars in distant galaxies, geneticists are better than ever at detecting new sequence changes in reproductive cells. The technical name for these errors is "de novo germline mutations." (The Latin phrase *de novo* is roughly translated as "from the new.")

Earlier in this book, we met patients with health conditions caused by new mutations—people like Astrea, who had a heart condition that almost claimed her life, or the pregnant woman with the blood disorder called paroxysmal nocturnal hemoglobinuria who passed away. But in each of these cases, the new mutation happened during development and affected only *some* of their cells. The acquired genetic conditions described in the preceding chapters have affected parts of the brain, lymph nodes, and skin, just to name a few. Mutations that strike only in those tissues can disrupt the life of the affected individual but are not transmitted to their children. In contrast, when the *reproductive* cells that go on to form an individual are struck with a new genetic error, it can get endlessly passed down to subsequent generations. In this way, a noninherited mutation becomes an inherited one. And since the genomes of the sperm and egg are recapitulated in every cell of a developing embryo, *all* the child's cells will carry that mutation. The stakes are extremely high.

There's a scientific awakening among geneticists that there are more genetic changes happening within our testes and ovaries while we live than previously thought, and that these might have implications beyond our own health to affect the genetic well-being of our children.

Part of the reason that de novo germline mutations have been overlooked is that finding them used to be an insurmountable technical challenge. Detecting how often these tiny changes within the three billion letters (or subunits) of DNA occur in each generation has been likened to measuring the frequency of needles in haystacks. But that has slowly begun to change. New sequencing tools and computational methods have made it easier and cheaper to pick up small typos and structural variations in the genome, beyond the major chromosomal abnormalities that can be detected with a microscope. One analysis of genetic data from twenty-eight people across four generations of a Utah family revealed around 150 de novo genetic changes *per generation* detectable by modern technology. That number was about 50 percent higher than previous estimates.

It's not unheard-of now to encounter people whose illnesses are linked to a de novo germline mutation. New research has linked such mutations to aggressive cancers such as retinoblastoma, which originates in the eyes, and osteosarcoma, which starts in the bones. Akiva Zablocki, the head of the Hyper IgM Foundation, whom we met briefly in chapter 3, knows about de novo mutations and their health impact. His wife, who is the youngest of seven siblings, was the only one in her family to have the hyper IgM mutation, which means it was most likely a de novo change in either the sperm or egg that formed her. Because she was a carrier for the trait, she passed it on to their son (who, thankfully, received a successful bone marrow transplant for the immune disorder).

Of the tens of thousands of de novo germline mutations identified in disease so far, almost half have been linked to autism (which might be a reflection of the robust research funding allocated to study this neurodevelopmental condition). A survey of forty-six thousand mutations found that 45 percent had been associated with autism. This strong signal correlating genetic changes in reproductive cells with autism is notable because prior research has suggested that children born to older fathers, whose sperm is more likely to contain new mutations, are at heightened risk of this disease.

There's a hope that knowing more about de novo germline genetic errors will help scientists better understand the molecular machinery that breaks down in certain diseases—and that this, in turn, will point drug developers toward the cellular functions to target and resuscitate with new medications. Sometimes, discovering a de novo mutation can even provide long-sought answers about tragedies that seem to strike out of the blue: In recent years, doctors have identified cases in which these genetic aberrations were likely the cause of some previously unexplained sudden infant deaths. The reality is that congenital genetic diseases do not always go back millennia. Sometimes they creep in and cause unthinkable heartbreak within the time span of our lives.

■ ■ ■

The seemingly random genetic errors that sneak into the genome and upend lives are cruel and unfair. And when it comes to de novo germline changes, the playing field might be more uneven than we previously imagined.

Raheleh Rahbari has had a long interest in understanding the way genetic traits are passed on. When Rahbari was a child, she spent endless hours plucking books out of the library stacks of Iran's prestigious Sharif University of Technology, where her mother worked as a librarian specializing in biotechnology and biochemistry. "I had read during childhood about Mendel's experiments with peas," Rahbari says, referring to the nineteenth-century monk Gregor Mendel's research on heredity. Later, as a student in England, she learned how bits of DNA called transposable elements could essentially jump around the genome, including in sperm and egg cells. Rahbari became curious about how else these reproductive cells might go genetically haywire: "During my PhD, I had more time to think, 'Okay, so what other sort of things are going on in the germ line?'"

Rahbari eventually joined an effort at the Wellcome Sanger Institute led by its current director, Matthew Hurles, to parse information in one of the biggest genetic databases in the world. The database, administered by Genomics England, a government-owned company in the UK, had by then amassed DNA sequences from thirteen thousand parents and their children.

The analysis suggested that some people might be more prone than others to racking up genetic errors in their reproductive cells. More specifically, by reviewing the genetic data, the study indicated that men with faulty DNA repair genes had excess mutations in their sperm that they passed on to their children as germline de novo changes. Rahbari recalls when she and her colleagues had this insight. "We realized, 'Oh, there are some families—a very small number of them—that are hypermutators,'" she says. "They had a considerably higher number of de novo germline mutations than expected."

Rahbari also joined forces with a wide team, co-led by geneticists Claire Palles and Ian Tomlinson, that gathered more evidence for this kind of phenomenon. Their study took a closer look at around half a dozen families with known mutations in genes for DNA proofreading or repair, as well as several families without these variants who served as controls. The children of parents with defective DNA repair had extra de novo mutations that appeared in regions of the genome unlikely to produce any consequence. But the scale of mutations was noteworthy. By the scientists' calculations, parental DNA repair defects were responsible for about 20 to 150 more new germline mutations per child than those seen in children within the control group.

It's not just inborn predispositions that tip cells toward accumulating more de novo mutations that get passed down. External forces can add these changes, too. This had come up when the Wellcome Sanger group analyzed the giant database set up by Genomics England. Within that pool, there was genetic data from thousands of families affected by various diseases. Digging around, the researchers found something unexpected. In several children they found mutational patterns associated with chemotherapy—but the baffling twist was that these youngsters had never had cancer. "That was very, very strange because these children, for instance, were never exposed to chemotherapy reagents, but we could see the signature of mutations that are usually caused by exposure to cisplatin, which is a very well-known chemotherapy [drug]," Rahbari says. "It was very striking, and we found this very confusing."

The team went back and checked the medical records of the children's parents. Then the answer hit them: "We realized that the fathers of these kids actually received cisplatin prior to conceiving their children," Rahbari explains. The men had undergone treatment for cancer. Usually, men diagnosed with cancer will bank sperm samples as a sort of insurance policy in case the treatment they are about to undergo compromises their fertility. But the fathers of the children with the cisplatin mutations had not used stored sperm samples to start a family—perhaps because they had not been advised to do so. "I found it quite astonishing in a way," Rahbari says. Unfortunately, the damage to their sperm from chemicals was reverberating in the next generation. It's unknown what health effects chemotherapy-related mutations will produce when they are passed on, but there's a worry that they could possibly raise the children's lifetime risk of cancer and other diseases. It's a stark example of why we need to know more about which drugs and environmental exposures might stir up new errors in our reproductive cells with the potential to harm future generations.

= = =

Most of the time, when doctors talk about the risk of genetic problems in parents' reproductive cells, they focus on egg cells. Women are urged not to delay motherhood—not only to ensure they don't miss their fertile years, but also to ensure the good health of any children they might have. Doctors often note that the risk of conceiving a child with chromosomal abnormalities increases exponentially as a woman gets older. Take for example Down syndrome, the most common chromosomal disorder. Around one in eight hundred babies born in the United States have this condition, in which an extra chromosome (usually a result of a cell division error in the ovaries) causes developmental and health issues. Studies have shown that maternal age has a huge influence on the risk of this malady. Before widespread prenatal genetic testing, less than one in one thousand babies born to women in their mid-twenties had Down syndrome, compared with one out of every fourteen for women in their late forties. Having an abnormal number of chromosomes is strongly linked to defects in eggs that occur with so-called "advanced maternal age."

In recent decades, however, a small group of scientists has started to challenge the medical narrative that the bulk of new genetic abnormalities inherited from parents trace back to egg cells.

When researchers started applying technologies looking at smaller-scale de novo germline mutations in children, they saw that the majority of these tinier genetic errors did not trace back to mothers. In other words, the advent of better DNA sequencing has also expanded the focus of research into transmitted mutations beyond eggs to include sperm. The greater risk of genetic disorders from older dads—known as the "paternal age effect"—has been associated with a growing list of diseases. Several of the disorders linked to this phenomenon, such as Apert syndrome, Crouzon syndrome, and Pfeiffer syndrome, involve serious facial or skeletal abnormalities.

Rahbari's work measuring the extent of de novo germline mutations has opened her eyes to how much these changes can matter in sperm. When she began in this area of research, there wasn't a good method to study the germ line directly. Instead, she and her colleagues studied family trios—each made up of two parents and their child—and compared the DNA from each kid with that from their mom and dad. They found that many more mutations came from fathers than from mothers. Rahbari laments that this message hasn't yet sunk in for the public—and even many doctors. "Mom always gets the blame," Rahbari says. "In fact, about 80 percent of germline de novo mutations are from the fathers."

The findings from Rahbari and her colleagues were echoed a couple of years later by a giant study in Iceland. A DNA analysis in Iceland of fifteen hundred family trios (each consisting of a child and two parents) indicated that children inherit four times as many new mutations from their dads than their moms. The numbers suggested, for example, that a baby will receive forty-five new mutations from a thirty-year-old father compared with only eleven from a thirty-year-old mother.

Part of what makes sperm so prone to errors is the seemingly endless replication involved in their production. The stem cells that produce sperm divide continuously and in great numbers. Each pair of testes makes millions upon millions of new sperm a day. By some calculations this ends up being as many as two thousand sperm produced with every heartbeat. In addition to going through replication cycles—where there is a chance for DNA errors to be introduced—sperm are well documented to be vulnerable to damaging free radical molecules, and to genetic disruption by harmful chemicals. The sperm of a twenty-five-year-old man is the end result of around 350 cell divisions tracing back to precursor stem cells. The sperm of a forty-five-year-old will have undergone 750 cell divisions. Perhaps it's no surprise, then, that older fathers pass on more mutations to their children than young dads do. One study of germline mutations in Icelandic families found that children born to forty-year-old fathers had twice as many new mutations passed on to them as those with twenty-year-old dads. These sorts of discoveries are of particular interest because the age of fathers at the moment of conception is trending higher and higher.

Men typically produce sperm from puberty until their last breath. Sperm production depends upon the hard work of a dedicated population of cells. These stem cells can acquire mutations, most of which are inconsequential. But some of these sperm-producing cells may pick up genetic changes that give them a leg up. Those that do might gain a survival advantage and gradually outnumber their counterparts as a person gets older. This has been dubbed "selfish selection." If these corrupted stem cells start winning out, then a greater proportion of sperm will be produced by them, which of course carry the same DNA mutations. (The process is reminiscent of the phenomenon described in chapter 4 called clonal hematopoiesis, in which mutant stem cells start to dominate blood production in the body.)

The selfish selection of stem cells in the testes reminds us of the small-scale Darwinian battles that can and do play out within our bodies. But in this instance, the potential influence of the changes in DNA extends beyond the individual in which they are happening. This is what makes de novo germline mutations so important. Some of the mutations identified in spermatogonial stem cells that selfishly start taking over might have potential health consequences for future generations. They include mutations seen in cellular pathways linked to cancer and to a genetic disorder called Noonan syndrome. A mosaic of sperm with different genetic identities can exist within men's testes. And although this phenomenon doesn't affect the health of dads, it can affect their kids. "While sperm mosaicism has few consequences for men," a group of scientists wrote several years ago, "the offspring and future generations are unwitting recipients of gonadal cell mutations, often yielding severe disease."

There's a lot still to learn about the environmental factors that might nudge sperm and their precursor cells toward genetic change. There are hints that smoking might increase mutations in sperm, for example. On the flip side, one analysis suggested that Amish individuals show a small but significantly lower rate of new mutations—around 7 percent less—than other populations, even after controlling for the fact that they have younger parents. The scientists behind that research note that the Amish limit the use of technology and might be exposed to fewer mutagens. It's very early days for this type of research, and it's hard to say for sure what influences the rate of de novo mutations.

In the meantime, some scientists have begun to test out the possibility of screening embryos for de novo mutations that originate in sperm. They sequenced DNA of the sperm of three men who had undergone fertility treatment with female partners. The scientists, who included Joseph Gleeson of the University of California, San Diego, compared that genetic data with that from the men's blood and saliva. This comparison allowed them to determine which mutations had arisen uniquely in the men's sperm. The group identified a total of fifty-five new mutations unique to the sperm. Notably, not all of the men's sperm contained these mutations, so the next step was to see whether those that did possess them had successfully created embryos. Among the embryos tested from the couples, nineteen of them carried de novo germline mutations transmitted to them by sperm. That result demonstrates that the ability to screen embryos for de novo germline mutations is within grasp, although the *feasibility* of doing so on a grand scale is still out of reach, and it would be nearly impossible to detect all de novo mutations that might be passed on. Despite the questions about feasibility, the possibility of having this sort of power to select against new mutations is radically new and thought-provoking.

■ ■ ■

It's abundantly clear that mutations can strike reproductive cells as well as other cells throughout the body. But part of the mission of scientists such as Rahbari is to decipher whether there is something unique about how mutational processes play out in sperm and eggs. These cells carry a lot of responsibility, and she and others have wanted to learn whether they have any tricks to fend off threats to the integrity of their genomes.

Rahbari wondered how reproductive cells compared with those of other tissues. Were they just as prone to racking up DNA mutations as, say, cells in the skin or airways or elsewhere? For many years, scientists speculated that germline cells had different rates of mutation than regular cells. But they based that assertion on indirect evidence. No one had shown it directly. So Rahbari helped oversee a massive collaboration across the institute where she worked, as well as with scientists from the United States, the Netherlands, and other countries. The team used lasers to dissect extremely tiny portions of tissue from several deceased donors, ranging from the liver to the appendix and the thyroid to the testes. They then sequenced the DNA of these samples separately and compared twenty-nine different cell types. The highest rates of single-letter DNA changes—around fifty or so a year—seemed to take place in tiny folds in the lining of the small and large intestines, as well as the appendix.

The number of such mutations in the cells that *generate* sperm was a drop in the bucket by comparison: They seemed to acquire as few as two or three a year. Those sperm stem cells had a rate of mutation accumulation twentyfold lower than cells in some nonreproductive tissues from the same individual. Rahbari was shocked at the extent of their genetic integrity. "To me, it is still astonishing because all of the intrinsic and extrinsic factors are the same in the colon versus the testes, but somehow [cells in] the testes managed to have a lot less mutation," Rahbari says. "I think it is quite remarkable."

It wasn't feasible for Rahbari and her colleagues to study the mutation rate in eggs in that study, but there's plenty of interest in how such cells in ovaries might guard against mutation.

Consider for a moment the precarious situation of an egg cell. Female mammals are born with a reserve of immature eggs that begin to be used once they go through puberty. Although some scientists believe stem cells persist in the ovaries to replenish this stock, there is a huge amount of debate over this. The stock of eggs a female is born with is—according to the traditional scientific dogma—all they will ever have for the duration of their lives. The eggs in reserve need a way of remaining free from damage until the animal has completed their reproductive life. In humans, this can take more than forty years. It's a tall order to last that long, but egg cells are up to the task. To make it to the finish line, they have several tricks up their sleeves.

Egg-producing cells are sleeping beauties. They stay for many years in an immature state, arrested in one of the stages of cell division, and wait until they are called into action. Whereas precursor cells for sperm undergo an estimated thirty cell divisions before puberty and continuous further divisions to generate little swimmers, those that produce eggs go through around twenty-two cell divisions in the womb and remain in an arrested state thereafter. Only *one* further cell division initiates in immature eggs, and this happens only if they are prompted to advance in the final stages of ovulation.

One person who has devoted much of her career to deciphering the specialness of egg cells is Kateryna Makova. Like Rahbari, Makova's path toward studying biology was influenced by her childhood. Makova grew up in Ukraine and was there when the Chernobyl nuclear accident occurred in 1986, just around sixty miles away from her home city of Kyiv. "I was a teenager when the Chernobyl disaster happened, and we all were wondering what effects it might have on our health," Makova recalls. "As a high school student, I expressed some interest in genetics. I always had this question in the back of my mind about whether there was an elevated mutation rate in the organisms who experienced Chernobyl—including humans and other mammals." Makova continued pursuing a path in science and ended up living in a town just outside Chernobyl for several weeks during graduate school so she could collect samples from animals, mostly voles, to study. For her dissertation, she documented the genetic consequences of the nuclear accident.

When Makova later got a job as a scientist at Penn State University, she decided to study bean-shaped structures within the cell called mitochondria. Mitochondria are known as the batteries of the cell because they produce energy for important biological reactions inside. Crucially, mitochondria possess their own genetic material, which is separate from the DNA that makes up the human genome inside the cell nucleus. Their genetic repertoire is also much smaller. Mitochondrial DNA contains only about sixteen thousand letters in its sequence, whereas the human genome consists of more than three billion. Makova had also probed mitochondrial DNA during her time as a researcher in Chernobyl. What appealed to her was its abundance of mutations. Mitochondrial DNA has a mutation rate about an order of magnitude higher than the DNA of our genome (perhaps because of the damaging molecules, known as free radicals, that are rife within mitochondria). More mutation means more anomalies to analyze.

Makova has—like Rahbari—embarked on recent studies comparing the mutation rate of reproductive cells with those in other tissues. But rather than look at the genetic material in the nucleus of those cells, she has focused on changes in their mitochondrial DNA. For part of her research, Makova decided to try collecting tissue from macaque monkeys as a proxy for what might be happening in humans.

Some of the specimens came from macaques undergoing necessary abdominal surgery, others came from macaques that had died naturally. The goal was to avoid sacrificing any monkeys for the project. As a result, it took several years for Makova's team to collect enough samples to run their study. Each time a sample became available, there was a scramble to get hold of it. The need for speed and delicacy was particularly true for the ovaries. After excision, they had to be shipped off within hours at 37 degrees Celsius (body temperature) in a toaster-oven-sized, battery-operated incubator so that the scientists could harvest the macaque egg cells from within them for study. "Each of these was a separate FedEx shipment," Makova says, reminiscing about the cost and effort involved. "It was not cheap."

The Makova-led experiment showed, like prior studies, that reproductive cells mutated more sparingly than other cell types. Of all the categories studied, egg cells accumulated mitochondrial DNA mistakes most slowly, and cells in the metabolically active liver racked up these errors fastest. There was something else interesting about the egg cells of the female macaques. Not only did they seem the least prone to acquiring genetic faults; they also seemed to stop accumulating as many mitochondrial mutations after the animals, who undergo puberty around age three, had matured. "The mutation frequency increased slightly until the age of about nine years. And then it plateaued," Makova says. "This is not the story that we were expecting. But, you know, we cannot make it up. It is what it is. We double-checked the data multiple times." She speculates that a protective mechanism somehow prevents the accumulation of mutations in egg cells. More recently, her team studied samples from a group of women twenty to forty-two years of age and found a correlation between higher age and more mitochondrial mutations in blood and saliva but not in eggs, which don't seem to rack up as many of these changes.

One way egg cells might preserve themselves is by staying low-key: A group of researchers in Spain have evidence that eggs keep their genomes from getting mucked up by altering their energy production. All the cells in the body—including eggs—get energy from the movement of electrons in their mitochondria. Sometimes the electrons in the mitochondria leak and produce the previously mentioned free radicals, highly reactive molecules and atoms that attack and disrupt DNA. It turns out that eggs have tamped down the process that generates free radicals. The team in Spain, led by Elvan Böke at the Centre for Genomic Regulation in Barcelona, found this when they studied egg cells from the African clawed frog and from humans. While in their resting state, the cells seemed to go without one of the active complexes involved in shuffling electrons for energy in mitochondria. The result? They had lower energy production—and likely fewer free radicals.

Makova says that the findings could perhaps explain how mitochondrial DNA mutation rates are relatively low in egg cells, as she found in macaques. "It goes very nicely with our findings. We were happy to see it," she says. She believes more research is needed, however, and that there might be other, unknown mechanisms that protect egg cells from DNA errors.

Another way egg cells seem to maintain their genetic integrity is by being good at repair. When Australian scientists exposed female mice to gamma rays, they found some evidence that the rodents' immature eggs had cellular mechanisms to correct harm from that radiation. More specifically, the eggs were highly capable of fixing double-strand breaks in their DNA, which are among the worst kinds of genetic damage a cell can experience.

Even as geneticists have learned about the ways that egg cells seem even less likely than sperm cells to end up with new mutations, many doctors haven't heard the message. "You know what scares me? Some clinicians still think that maternal age is the main contributor to germline de novo mutations," Rahbari says. "I was hoping our publications might change people's perception, but clearly it might have only been read by scientists."

It's clear that egg-forming cells have the potential to teach us something about how to skirt mutations. As much as it has become evident that our reproductive cells acquire numerous mutations throughout our lives, they seem to do this at a slower rate than other cells in the body. "The stunning part is how the germ line does such an amazing job protecting the cells," Rahbari explains. "Just saying that the germ line protects itself because of evolutionary rules—that's easy to say. But how does it do that? I think it could be very important biology to uncover."

A greater understanding of how reproductive cells fend off mistakes in their DNA could be medically useful. If we pinpoint the mechanisms, we could perhaps ramp them up if we want to stave off mutations, including in nonreproductive tissues. As we will see, attempts to keep genetic mistakes at bay with drugs have already begun. But it turns out that scientists are not only preoccupied with the mutation of our own cells. They're also trying to slow the mutation of infectious life-forms that reside within us and that use us as their unwitting hosts.

Strains in the System

The stomach is an unwelcoming place in our bodies. The cells in its lining excrete gastric secretions that are more acidic than even lemon juice or vinegar. It's hard to imagine that any living thing could survive there for long. But if you thought it was free from microbial invaders, you'd be mistaken. One aggravating bacterium that excels at making the stomach its home is *Helicobacter pylori*. Globally, around four in ten people carry or have carried *H. pylori* in their stomachs. The proportion varies geographically—somewhere around 80 percent of adults in Russia are thought to have been infected with it at some point in their lives, whereas that number in the United States is perhaps closer to 18 percent—but it's definitely pervasive. During the first three years of the pandemic I was able to avoid getting Covid, but I was not able to dodge falling sick from *H. pylori*, and it was a terrible and painful ordeal. I could barely eat and lost twenty pounds within a matter of months. Others who are infected have it much worse.

H. pylori's history is deeply intertwined with our own. It's been infecting our species at least since a group of humans migrated out of Africa around sixty thousand years ago. Even the 5,300-year-old ice mummy nicknamed Ötzi, discovered by hikers exploring the Tyrolean Alps in Italy, carried *H. pylori*.

Then there's Charles Darwin. Darwin was only twenty-seven years old when he returned from the voyage that inspired his theory of evolution, but his health would never be quite the same. For the next several decades, he suffered from long bouts of severe tiredness, skin problems, headaches, vomiting, and gut pain. The letters Darwin wrote to his close friend, the famed botanist Joseph Dalton Hooker, reveal how much his stomach troubled him. "I have had a bad spell. Vomiting every day for eleven days, and some days after every meal," Darwin wrote to Hooker in 1863, at age fifty-four. Several months later, he wrote Hooker again about his health woes, saying, "I seldom throw up food, only acid & morbid secretion; otherwise I [should] have been dead, for during more than a month I vomited after every meal & several times most nights."

There has been no shortage of speculation about what ailed Darwin. Theories have ranged from arsenic poisoning to lactose intolerance to lupus, just to name a few, but it's very difficult to know for certain. At least a couple of doctors have speculated that somewhere along the way he picked up *H. pylori*. That, they say, is probably a leading factor behind his long-lasting gastric troubles. As evidence, they point to the fact that on the day he died, from what is thought to be a heart attack, Darwin had been retching and brought up blood—perhaps a sign that this bacterium had ravaged his gastric lining.

One person who has posited that *H. pylori* was behind Darwin's tummy troubles is Australian gastroenterologist Barry Marshall. According to Marshall, these days a patient with Darwin's symptoms would almost certainly receive blood tests or an endoscopy to look for signs of *H. pylori*.

Marshall has a long personal history with *H. pylori*. As a young clinical fellow, he had become interested in the work of a fellow Australian named Robin Warren, who had identified a small corkscrew-shaped bacterium—which would become known as *H. pylori*—in the stomach biopsies of many patients. Warren, a pathologist, had also noticed that there was usually inflammation associated with its presence. Marshall succeeded in isolating and growing this bacterial species in the lab in the early 1980s. He felt it was responsible for stomach ulcers—but this went against the dogma at the time that such ulcers were due to psychological stress.

Marshall grew frustrated. He believed that antibiotics could be a much-needed solution to ulcers by ridding the stomach of the corkscrew bacteria. But the microbe infects only primates naturally, and he couldn't get it to do so experimentally in pigs, mice, or rats. "It was desperate: I saw people who were almost dying from bleeding ulcers, and I knew all they needed was some antibiotics, but they weren't my patients," he recalled in a *Discover* magazine interview years later. "So a patient would sit there bleeding away, taking the acid blockers, and the next morning the bed would be empty. I would ask, 'Where did he go?' He's in the surgical ward; he's had his stomach removed."

Faced with this maddening situation, Marshall took a drastic step. He decided to prove the danger of the corkscrew bacteria by experimenting on himself. First, he underwent a baseline gastric biopsy to show that he didn't have the bacteria in his stomach to start with. Next, he mixed up a solution that contained *H. pylori* microbes from the lab, which originally came from a patient with gastritis. He swizzled the organisms around in a cloudy broth and drank it. "My stomach gurgled, and after five days I started waking up in the morning saying, 'Oh, I don't feel good,' and I'd run in the bathroom and vomit," he later recounted to *Discover*. He was able to go to work, but felt awful.

Marshall was tired and not sleeping well. His colleagues told him he had "putrid" breath. "After 10 days I had an endoscopy that showed the bacteria were everywhere," he said. "There was all this inflammation, and gastritis had developed. That's when I told my wife." Marshall was able to take antibiotics after infecting himself with the bacterium and recuperated. And the work helped lead to the understanding that *H. pylori*, rather than stress, was the leading cause of stomach and intestinal ulcers. Marshall and Warren would ultimately share a Nobel Prize for their insight. But *H. pylori*, it turns out, is linked to more than just stomach ulcers.

■ ■ ■

Around the time that Marshall was infecting and curing himself of *H. pylori*, Nina Salama was growing up in the Midwest. As a child, she would hear many stories about different bacteria from her father at the family dinner table. He worked as a pathologist, a job that required reviewing the slides of biological samples taken from patients to find or confirm the causes of what ailed them. One evening in the mid-1980s, when Salama was a teenager in high school, her father shared an exciting development. He had heard about a newly identified kind of bacterium. The microbe had a corkscrew shape and appeared to cause ulcers. Salama's dad felt so fascinated with the discovery that he had decided to take a second look at the stored samples in his pathology lab. "He had gone back and looked in his slides to see these spiral organisms," Salama says. Lo and behold, he found *H. pylori* in the slides. It had been there all along—and he was so amazed that he told his family.

A decade later, Salama was interviewing for a job in the laboratory of the famous Stanford microbiologist Stanley Falkow. Had she heard about *H. pylori*, Falkow wanted to know. She told him she already knew about it. "I said, 'That's the ulcer bug,'" she recalls. Then Falkow gave her a preview document describing how the World Health Organization was going to declare *H. pylori* a carcinogen. "I didn't know about that cancer connection until Stan shared that with me. And I was like, 'Wait, a bacterium causes cancer?'" It came as a shock, Salama says. "I was kind of floored." Salama started on Falkow's team in 1995, and she has been working on *H. pylori* ever since.

In the early 2000s, Salama relocated to the Fred Hutchinson Cancer Research Center in Seattle and started her own lab. Around that time, she was part of a scientific team behind a key paper about how *H. pylori* can evolve inside a person. The study had examined *H. pylori* in a forty-eight-year-old Tennessee man who had refused antibiotics. They were told that he had turned down the medications because he had not had a good experience with such drugs in the past. Because he had not taken any therapies, the scientists could see how *H. pylori* mutated in the absence of treatment. They analyzed stomach biopsies from the patient taken six years apart. The genetic changes they saw in the bacteria's sequences from the different time points led them to conclude that *H. pylori* has a remarkable capacity to lose—and possibly acquire—DNA over time. It suggested to the research team that *H. pylori* undergoes "continuous microevolution" within its host.

Some twenty years later, Salama and her fellow scientists decided to take a closer look at the frozen bacteria that had been cultured from that Tennessee man's samples. In the two decades since their earlier paper, a revolution had occurred in genetic sequencing. Researchers had a newfound ability to sequence the DNA of microbes like *H. pylori* in exquisite detail, right down to the single-letter changes in its code. Salama wanted to apply these tools to the man's stored samples.

The reanalysis revealed more detail about the sophisticated ways in which *H. pylori* bacteria had evolved within the man during the six-year gap between his stomach biopsies. For one thing, the *H. pylori* microbes had acquired mutations that seemed to heighten their ability to cause inflammation. This is significant because scientists suspect that when *H. pylori* takes up long-term residence in the gut, it eventually creates enough chronic inflammation to tip the tissue there toward cancer.

The other key finding from the reanalysis was that the *H. pylori* bacteria were shape-shifting. Some had begun to lose their typically helical form and become more rodlike, while others had developed more tightly wound helices. In their 2020 paper describing the result, Salama and her colleagues speculate that such shape-shifting gives *H. pylori* a survival advantage. A follow-up paper from the group a few years later showed that, compared with bacterial isolates taken from the early stage of the Tennessee man's infection, the more tightly wound isolates from the later stage of his illness were more adept at colonizing mice with diseased stomachs.

Salama says that the morphing of *H. pylori* in the stomach is particularly important to grasp. "Basically, it's changing the way that it can bind to the tissues," Salama explained to me. "So figuring out which of those variants are important for interacting with the more diseased stomach becomes an interesting way to perhaps intervene." She notes that doctors might also one day be able to use telltale genetic variations in *H. pylori*'s sequence as a biomarker indicating disease progression. For now, measuring the mutational propensities of *H. pylori*, and other microbes that infect us, is keeping researchers very busy.

. . .

You might be doing a bit of math in your head by this point, and the numbers might not seem to add up. If almost half of the world population has been infected with *H. pylori*, then why don't more people have stomach ulcers or stomach cancer? The answer is that there are probably numerous environmental and innate factors that explain why the bacterium stirs up big trouble only in some individuals. Another reason is that certain strains appear to be more damaging than others. On top of all that, as research by Nina Salama and others has suggested, the way that *H. pylori* mutates once within a host might have a large role in how things play out. And *H. pylori* is a king of mutation among bacteria.

Strange things can happen with *H. pylori*. In the 1990s, a group of researchers examined biopsies from a Lithuanian teenager and found that at least two different strains of *H. pylori* inhabited his digestive system. The discovery helped show that people often harbor multiple strains of *H. pylori* at the same time. They also determined that the strains of *H. pylori* had exchanged genes within him on at least six different occasions. It was the first direct evidence that *H. pylori* could recombine its genetic code inside an individual infected with multiple strains. This tendency of *H. pylori* to swap its DNA—a process known as "horizontal gene transfer"—is one of the reasons it has a higher propensity for genetic diversity than most other bacterial species. (Later studies have estimated that when multiple strains of *H. pylori* exist within a person, the rate of certain kinds of DNA swapping within the bacteria's genomes can accelerate their evolution by a hundredfold.)

It's no coincidence that *H. pylori* was the first bacteria of which multiple strains were genetically sequenced. The sequencing, which was published in the late 1990s, gave microbiologists a better reference point from which they could survey the genetic instability and small-scale evolution of the pathogen. It also revealed that *H. pylori* lacks multiple genes involved in repairing DNA errors that are found in other bacteria. Moreover, scientists began to appreciate that because *H. pylori* has a long infection time, over the years that it can reside in the stomach it has plenty of time to rack up new mutations.

The genetic insights only increased from that point forward. A paper that counted Barry Marshall among its authors detailed what happened when volunteers drank a beef broth containing *H. pylori*. The study revealed that there was a flurry of mutations in the microbe during the period immediately after the participants were infected. The propensity of those initial genetic changes to affect the outer membrane of *H. pylori* hinted at an evolutionary arms race between the bacterium and the infected person's immune system. The authors of the paper also noted that the flurry of mutations in *H. pylori* during acute infection is "orders of magnitude faster than mutation rates in any other bacteria."

Each year within a person, *H. pylori* accumulates around thirty new changes to individual letters, or subunits, of its genome—which is an astronomical amount, according to microbiologists. But other bacteria have a propensity to mutate as well. *Klebsiella pneumoniae*—the most common cause of hospital-acquired pneumonia—amasses around ten such mutations annually inside a host, according to some calculations, while the life-threatening intestinal invader *Clostridioides difficile* picks up about two per year. Meanwhile, a phenomenon known as hypermutation—which, as you might guess from the name, means more genetic alterations—was shown to fuel diversity in a sometimes-fatal bacterium known as *Burkholderia dolosa* within a cystic fibrosis patient.

The mutation happening in bacterial invaders can, unfortunately, bolster their resistance to drugs meant to eradicate them from the body. Look no further than tuberculosis, which is thought to be the leading cause of death from an infectious disease worldwide. The dangerous adaptability of the microbe that causes it—*Mycobacterium tuberculosis*—became apparent when doctors tracked it in a patient. The individual harbored a drug-sensitive version at the beginning, but over the course of three and a half years, during which time they progressively received seven different medications for the infection, it evolved to become "extensively" resistant to treatment.

H. pylori can also mutate to become impervious to certain antibiotics. There's evidence that treatments for it can create a population bottleneck within an individual—essentially a culling of the drug-sensitive bacteria that leaves the resistant ones to survive and then surge. And this isn't just a problem within the person in whom such mutations occur. Like many other pathogenic bacteria, *H. pylori* strains have been growing increasingly resistant to drugs worldwide. In the 1980s, between 0 and 9 percent of sampled *H. pylori* bacteria were resistant to the antibiotic called clarithromycin. Now, upwards of 20 percent of *H. pylori* samples in many countries are resistant to that compound. When I got sick with this microbe, I received a combination of two different heavy-duty antibiotics at the same time, as many patients do, in part to reduce the risk that it would evolve resistance within me. Some people with *H. pylori* have to take as many as four medications simultaneously.

■ ■ ■

It's not just *dangerous* bacteria that evolve within us. The beneficial ones we carry can also change over time.

As adults, each of our bodies is made of thirty trillion to forty trillion human cells, but we are probably host to a similar number of tinier microbial cells. The calculations have varied wildly among scientists, but some of them have suggested that a single person carries around thirty-nine trillion microbial cells within them. The human body is a landscape where microbial species mutate endlessly and sometimes vie for resources. It may sound icky, but many of them play an essential role in maintaining our health. Among the many ways they support us, the friendly microbes within the human gut microbiome help us break down our meals and absorb vital nutrients. It's a sublime symbiosis happening right inside our bellies.

Just before we enter this world, while still in the womb, our guts are essentially sterile. During the first years of life, a child will experience a rapid colonization and turnover of bacterial strains within the gut. One study of hundreds of infants suggested that their microbiomes shifted at a rate tenfold higher than in grown-ups. Anyone who has changed the diapers of a child can attest to the dramatic—and odorous—digestive transformation that happens in the first couple of years. Finally, by age two, the toddler will possess a stable, adultlike microbiome. And if the gods take pity on you, the child might even be potty-trained by that time. The youngster's microbiome evolution, however, is far from over.

As the years go by, the bacterial populations inside our bellies ebb and flow. Studies of the human digestive system have uncovered that some mutations help microbes gain a footing whereas other genetic changes cause some to fall behind. Microbiologist Isabel Gordo, who is based at the Gulbenkian Institute for Molecular Medicine in Portugal, wrote in an article that there is much more to learn about the "significant evolutionary dynamics" inside the gut. She added that recent findings "may only be the tip of the iceberg."

Although Darwin formed his concept of natural selection after traveling the world and visiting the Galápagos islands, we don't need to go such distances today to observe evolution at work. "Imagine that Charles Darwin was here today and we gave him a population genetics textbook and a next generation sequencer," Gordo wrote. "Would he go on another trip on the Beagle, or stay at home and study his gut microbes to find the molecular signatures of his theory of natural selection?"

Modern-day researchers have uncovered various ways that different bacteria evolve within the inner world of our intestines. One fascinating example involves a species of microbe called *Bacteroides fragilis*, a common occupant of a healthy human gut that helps us digest fiber.

A 2019 study showed how adaptable *B. fragilis* is. A team at the Massachusetts Institute of Technology analyzed stool samples from a dozen people, some of whom gave several samples over the course of two years. The team sequenced the genomes of more than six hundred *B. fragilis* isolates from thirty fecal samples in total. The analysis revealed that at least sixteen of the microbe's genes mutated within the participants. Eric Alm, the MIT professor who oversaw the study, expressed his amazement about the findings at the time. "The strains of *B. fragilis* that are growing in humans have been in that gut-like environment for millions of years, so the idea that encountering a new host's gut would induce a bunch of new adaptive mutations, and that these commensals would still be rapidly evolving, was surprising to us," Alm said. He and his colleagues speculated that differences in personal diets might underlie the genetic changes they witnessed in *B. fragilis*.

Alm's team has also elucidated *how* microbial evolution happens in the human gut. Scientists previously knew that microbes can absorb DNA from dead neighbors or build tubes to connect and swap genetic code via horizontal gene transfer. Researchers such as Alm have found that DNA exchanges like these were happening in the human gut.

In one study, he and his collaborators looked at thousands of bacterial strains from fifteen different human populations. They found that there were higher transfer rates of antibiotic resistance genes in the gut microbiomes of nonindustrialized populations—perhaps linked to the higher environmental prevalence of such resistance genes in the regions where those groups live. In contrast, they found that people in industrialized nations had higher overall rates of horizontal gene transfer. Alm and his coauthors speculated that the diets of industrialized countries—which are abundant in processed food—might create an inflammatory environment in the intestine that favors species more prone to swapping DNA via horizontal gene transfer.

After learning about the ecosystem of microbes inside the gut, many people want to shape it in a way that benefits their health. If that's your goal, experts advise eating high-fiber foods. Some suggest aiming for at least thirty different plant foods a week. They point to research that followed sixteen hundred people for a decade. Participants who consumed the highest levels of fiber gained less weight over the ten-year study and had the highest levels of microbial diversity.

■ ■ ■

Some of the most consequential genetic changes that can happen within a person have to do with our viral inhabitants. Unlike a bacterium, a virus has no cell membrane, nor most of the stuff that usually goes inside. It's just a strand of either DNA or its cousin, RNA, sometimes simply packaged in a protein shell, that hijacks a host's cells and borrows the biological machinery there to replicate—and even mutate.

Proof that viruses don't stay genetically static inside their hosts came in the 1970s, when a scientist named Donna Sabo initiated an experiment in a laboratory at the University of Zurich. With the input and support of molecular biologist Charles Weissmann, she and the small research team took bacteria in the lab and exposed them to a virus known to infect the microbes. A remarkable genetic diversity emerged over time in the viral population. It was the first demonstration that viruses could produce new and different versions of themselves during an infection. The discovery was a turning point, marking the beginning of a whole new field of experimental studies of viral evolution. (Sadly, due to Sabo's untimely death, she didn't live to see the publication of the study's results.)

Within a decade, the world was hit by a pandemic that ultimately showed the urgency of understanding the ability of viruses to mutate within a host. The disease, AIDS, was caused by HIV, which is particularly adept at generating genetic shifts to sidestep drugs and the immune system. In 1995, about the time that new infections peaked, a clearer picture emerged of how badly medications aimed at the virus were failing. Even when researchers gave different medications to people with AIDS and reduced the size of their infections by 99 percent, the virus still rebounded within two weeks, by which time it was resistant to the treatments.

It turns out that when someone is initially infected with HIV, there is a massive replication burst inside them that produces staggering mutational diversity. In fact, there is ultimately as much diversity of variants in the person's body as there are in influenza virus samples worldwide in any given year. Without proper treatment, the body of an infected individual will produce millions of viral variants in a single day. Eventually, there are evolutionary bottlenecks, meaning that some

HIV variants will come to dominate. Some of these dominant viruses are then transmitted to the next person, in which a different bottleneck selects the variant that will drive the infection when passed to additional individuals. The cycle happens again and again—a burst of mutations, followed by another selection bottleneck.

HIV is far from the only virus that evolves within us. Researchers have found evidence of rapid mutations among the viruses that cause polio, Zika, dengue, and influenza, just to name a few. Essentially, once a virus starts replicating inside of you, your body becomes a breeding ground for new variants of the pathogen—which, as in any other ecosystem, may begin to compete against one another for dominance. Scientists are even exploring the markers of genetic change in bird flu, in spillover events in which humans are infected, to see if the virus adapts once it's inside a person.

During the Covid pandemic, the worry about the novel coronavirus's ability to diversify and develop new tricks became acute. Within the first year of the pandemic, a new variant of SARS-CoV-2 was linked to massive explosions in the number of cases, and to a higher risk of death. It appeared that this coronavirus had picked up powerful mutations that made it better at binding to entry sites of cells to infect them. It also seemed to evolve a way to evade the antibodies that protect against versions of SARS-CoV-2 that had been circulating earlier in the year. Meanwhile, other worrisome variants were cropping up around the world, in South Africa and Brazil.

It was becoming increasingly clear that one place the coronavirus liked to churn out new variants was inside our own bodies. Most people with Covid are thought to stop shedding infectious virus after around ten days from the time they were infected, but numerous outlier cases were being identified. These tended to involve people with weakened immune systems. New strains sometimes emerge in immuno-suppressed individuals, in part because of the unique evolutionary dynamics at play when they are unable to clear the coronavirus quickly from their bodies and it lives and evolves inside them for months.

In one case, reported in late 2020, Michigan researchers tracked the sequences of SARS-CoV-2 in a sixty-year-old man with cancer. The man was on medication to suppress his immune system's B cells, which normally produce antibodies, so he had a weakened ability to fight the virus. It lingered longer in him. Over the course of the four months that researchers tracked his SARS-CoV-2 infection, the virus's spike protein, which is the principal target of Covid vaccines, remained unchanged. However, the scientists found other mutations that popped up elsewhere in the virus, unrelated to the spike protein. This suggested that the virus was mutating inside him.

This was far from the only such example. An additional report also emerged about an immunocompromised seventy-one-year-old woman within whom the virus was also changing. The woman had presumably become infected at a rehabilitation facility in Washington state in February 2020 following a spinal surgery related to the cancer she was fighting. She ended up harboring the coronavirus for upwards of three months. She was deemed to be infectious for more than two-thirds of that time. The virologists analyzing the viral sequences she carried during this period "observed marked within-host genomic evolution of SARS-CoV-2 with continuous turnover of dominant viral variants." In other words, the virus was definitely evolving new strains—including some with changes to the spike protein sequence—during the course of her infection. No sooner had this study come out than another report appeared, this time in *The New England Journal of Medicine.* It described the tragic case of an immunocompromised forty-five-year-old man who was infected with replicating SARS-CoV-2 for many months, and succumbed after almost half a year. As in the case of the Washington woman, many of the mutations that evolved in the patient occurred in the virus's spike protein. This created concern that there was a potential for accelerated evolution of the coronavirus in a small subset of immunocompromised individuals with persistent infection.

The coronavirus mutates in healthy people with normal immune systems as well. One reason it is so prone to changing is that it is a master of remixing its genome through a process called recombination. Unlike small mutations, which are like tiny typos in a genetic sequence, recombination is akin to overwriting the second half of a sentence with a slightly different version. When that happens, versions of SARS-CoV-2 emerge with completely different chunks of sequence. As the months and years of the pandemic passed, scientists found more evidence that this phenomenon was influencing the emergence of new versions of the coronavirus. For example, the Omicron XE subvariant is thought to have emerged after a recombination of the BA.1 and BA.2 subvariants of Omicron, which were both circulating at the same time. Presumably, some people were simultaneously infected with both BA.1 and BA.2, during which the different subvariants recombined. The worry is that when we are infected with a harmful virus like SARS-CoV-2, the virus can sometimes become more and more dangerous as time passes. That evolution can cause a problem not just for the person within whom the virus is changing—it can pose a problem for everyone on the planet.

We're constantly trying to stay one step ahead of viruses with vaccines that can stop them from gaining a foothold in the body. And booster shots, as you saw in chapter 3, work to keep the evolutionary arms race in favor of our immune cells over foreign pathogens. Sometimes, in very rare examples, we even need to keep the good viruses that protect us from mutating.

. . .

In 2017, a few years before the global coronavirus pandemic hit, an experiment intended to fight another virus got off the ground. Within three days in April, builders combined prefabricated modules to construct a temporary sixty-six-unit container village—called Poliopolis—in a parking lot on the campus of the University of Antwerp in Belgium. The amenities included a lounge with a flat-screen TV and foosball table, along with a fitness room. The complex also had a common kitchen and dining room, along with a library where participants could read in peace.

Some of the smaller details hinted at the special nature of the site. The kitchen was equipped with a one-way glass window that permitted study volunteers to look outside but made it impossible for the public to see into the facility. An open-sky atrium area set up with garden furniture served as the place participants could lounge and even barbecue, since they weren't allowed to go off-site. When it was ready, thirty healthy adults entered the containment facility and remained isolated for twenty-eight days. Once inside, they received an important immunization: a dose of one of two newly designed vaccines against polio.

Polio, which can cause life-threatening paralysis, was once a global killer like Covid. Some people can still recall the iron lung chambers introduced in the 1930s to help affected individuals breathe. In the 1940s, polio paralyzed around thirty-five thousand people each year in the United States alone.

Thankfully, during the mid-twentieth century, a rapid succession of discoveries helped show how to shield against polio. In 1948, the immunologist Hilary Koprowski combined a sample of carefully weakened poliovirus with rat brains in a kitchen blender in his laboratory and macerated them together. He poured the greasy, cold mixture into a beaker and drank it. It tasted like cod liver oil. Koprowski's vaccine contained attenuated poliovirus, meaning that the pathogen had been weakened so that it could not make people sick but could still train the immune system to fight the virus. Koprowski refined his vaccine, and it was used in some parts of the world, but it didn't gain US approval and never really took off. More widely adopted prophylactics sprang out of the work of two other scientists. The virologist Jonas Salk developed a vaccine from inactivated polio, which was first used in 1955, and the microbiologist Albert Sabin created a live, attenuated oral polio vaccine that received a license from US regulators in 1961.

With both the Salk and Sabin vaccine options in hand, the world mounted a giant battle against polio. When the World Health Assembly convened at its annual meeting in Geneva in 1988, it resolved to eradicate polio from the globe. The attendees set a lofty goal—but it seemed within reach. Toward the end of the century, the number of polio cases had dropped by a whopping 99 percent. By early 2003, the virus was detected in circulation in only seven countries in the world. At that point, one of the three types of poliovirus—type 2—already appeared to be wiped off the planet.

The efforts relied in part on use of the live attenuated vaccine given orally. (Why not just use the inactivated version? The inactivated vaccine, which is the only kind currently used in the United States, is injected and carries no risk of disease to the person receiving it. However, it doesn't produce as robust intestinal immunity against poliovirus as the oral vaccine that originated from Sabin's lab.) The oral polio vaccine gives strong immunity, including in the gut, where the virus usually likes to enter the body. Moreover, it is given as liquid drops in the mouth—eliminating the need for sterile syringes and making it more convenient for mass vaccination campaigns.

Unfortunately, there was a blind spot.

One of the downsides of the oral polio vaccine is that sometimes the live virus it's made of can mutate to become paralytic again. This can happen within the person who receives the oral drop, and the active virus they shed when they go to the bathroom carries the same paralytic risk. The shed particles can enter the sewage system, and if that system is not well maintained, it can spread beyond there to infect other people in the local community who are unvaccinated. Despite the fact that *wild* polio type 2 has been quashed, the *vaccine-derived* version of type 2 can enter communities this way. Worryingly, of every two thousand people infected with the vaccine-derived version, roughly one will become paralyzed. Moreover, the outbreaks of this vaccine-derived polio have become more common in the last decade. Whereas only two cases were reported in 2016, that number jumped to 1,082 cases in 2020. (There was even a case of vaccine-derived polio in New York state in 2022, likely picked up in the United States from someone who had been overseas. It was the first polio case of any kind in the country in many years.)

Scientists, including those behind the Poliopolis project, have aimed to develop a better inoculation against type 2 poliovirus. They also needed to ensure that, unlike the attenuated type 2 poliovirus used in existing oral drops, the live virus in the new inoculation would not have a propensity to mutate back into a nefarious form. In other words, they needed to make a live-virus vaccine that wouldn't morph back into a paralyzing version.

Researchers knew that if they changed some of the code of the live, attenuated poliovirus deployed in the oral vaccine, they could make it less likely to mutate. To achieve this, they modified a region of its sequence to enhance its genetic stability. The likelihood it would revert to a harmful form plummeted because several evolutionary events would have to occur simultaneously for this to happen—and the probability of that is low. They also tweaked the virus's instructions for replication to make it less prone to remix its genes in the process. Once they had this experimental vaccine worked out, they began testing it in carefully constructed settings—including within Poliopolis.

The closed design of Poliopolis aimed to ensure that none of the vaccine strains that the participants received could infect the surrounding community. All wastewater from the site underwent decontamination. That included not only water from the toilets but also water from the showers, bathroom sinks, and washing machines.

The immune response generated in that experiment and further clinical trials in Panama—along with a low rate of reversion to a virulent version in the excreted virus—were so encouraging that global health agencies rallied to make one of the novel vaccines available. For the first time in half a century, health workers had a new poliovirus vaccine to add to their arsenal. They saw it as a key tool to stop outbreaks of vaccine-derived polio happening across the world. Rollout of the new polio vaccine began in March 2021, and within the next three years health workers had administered at least a billion doses of it across thirty-five countries. Scientists hoped that, thanks to careful genetic engineering, this new live poliovirus vaccine would have less propensity to mutate.

The story could end here, with a neat conclusion. Unfortunately, the new vaccine hasn't been the slam dunk that researchers had wished for. A couple of years after the rollout, the Global Polio Eradication Initiative reported that two outbreaks had occurred in places using it. Polio experts contend that this was still fewer outbreaks than would be expected with the predecessor vaccine, but it shows how hard it is to prevent the poliovirus from shifting inside a host.

We will always have to stay one step ahead of our microbial nemeses. Even before the news of the problems with the new polio vaccine, scientists had analyzed stored fecal samples from clinical trial participants in Bangladesh who had received the old version of the vaccine against type 2 poliovirus—the original one that had a propensity to revert to become harmful. The stored fecal samples proved to be a gold mine. By sequencing them, the team showed that the original live, attenuated virus used to vaccinate against type 2 polio was prone to more kinds of mutations than previously thought. This suggests it will be important to monitor the rollout of new, genetically engineered poliovirus vaccines against the type 2 virus to make sure they don't mutate in unforeseen hot spots where genetic change is more likely.

It's crucial to keep tabs on how genetic shifts alter the behavior of our microbial passengers, says Adam Lauring, a virologist and infectious disease physician at the University of Michigan Medical School and one of the researchers behind the aforementioned study: "Evolution is just so important."

Ultimately, we are hosts to all sorts of morphing organisms. Whether it's the bacteria in our gut, the pathogens that naturally infect us, or the live viruses in vaccines, all sorts of nonhuman entities can mutate within us. Genetic shifts happen inside us at many levels. And researchers are not just trying to stop live-virus vaccines from mutating within us. Some are exploring ways to stop mutation in our own cells as a way to prevent a whole gamut of problems—even including aging.

Can We Stop Mutating?

At a four-way intersection in Punta Gorda, Florida, on the sidewalk between parked cars and a traffic sign, you will find a waist-high water fountain covered in seafoam-colored tile that has not aged well. It's not the most appetizing place to have a sip: The presence of sulfates makes the well water smell like rotten eggs. And that's not the only reason a passerby might have second thoughts about imbibing. A large white sign that covers one side of the blocky fountain alerts passersby that the water contains excessive levels of radioactivity.

Although the fountain is not able to offer perfectly potable water, it is flowing with history. In its heyday during the mid-1900s, the fountain drew long lines and grew so popular that, according to some sources, the handle on its tap had to be replaced every six months. Among the faded tiles are ones that feature a ship with its mast high and white sails full of wind. The vessel is meant to evoke the voyages of Juan Ponce de León and his fabled search for rejuvenating waters. In the late 1800s, centuries after his travels, the people of Punta Gorda claimed that this humble well was, in fact, the long-sought Fountain of Youth.

Of course, it's not as simple as that. Much like with a well, the deeper you dig into the story of the Fountain of Youth, the trickier things get. Multiple places have laid claim to possessing this fountain. There's a site in St. Petersburg, Florida, that advertises this, and there's another one across the state in St. Augustine as well. And it's not just Florida. Along the road to the South Bimini Airport in the Bahamas, where Ponce de León sailed through in the 1500s, a well carved out of limestone by thousands of years of groundwater is marked by a bright green-and-yellow sign that says "Welcome to the FOUNTAIN OF YOUTH." And then there is the matter of whether Ponce de León was ever really looking for a magical fountain. It's more likely that he set out for the New World seeking to seize land, riches, and—most horribly—slaves. Historians have noted that no mention of a Fountain of Youth exists in Ponce's contracts or correspondences.

In our modern era, we have yet to find a magic way to truly turn back the clock. But we have learned that for some people the clock seems to run a bit faster. Aging seems to happen at an accelerated pace. That was the problem that Michael Prescott faced. Prescott, who worked as a civil engineer designing bridges in Tennessee, started experiencing health problems in his thirties. On the surface, he appeared to be in very good health and would spend time hunting on the weekends. But on his son's fourth birthday, he had his first heart attack. He had four more heart attacks before the end of the following year. His family history seemed to be haunting him. Prescott's mother and grandmother had both died of heart attacks in middle age. His condition baffled doctors, but they decided he needed a heart transplant. So, in 2001, he underwent the procedure at the Vanderbilt University Medical Center in Nashville. But Prescott's woes were far from over. A few years later, he needed a kidney transplant. "My dad still had not had a diagnosis," says his son, Carter, who is now nearly thirty himself. "He was seeing doctors regularly. He had basically his own team at Vanderbilt who were trying to figure out what was going on with him."

As time dragged on, Michael Prescott's symptoms became more outwardly visible. His skin began wrinkling like people decades older than him, and he was developing cataracts. By his early forties, Prescott looked like he was sixty. When he would attend baseball games with Carter, people would mistake him for his child's grandfather.

Prescott's medical team continued to search for the underlying cause of his ailments, but he refused to sit by the sidelines. He decided to try to find out his diagnosis by himself. He would sit for hours in the living room in his favorite recliner, his slim form enveloped in a sweatshirt displaying the logo of his favorite football team, the Tennessee Volunteers, as he read one research article after another. "He had a hard time sleeping at night," Carter recalls, "and so he'd be in his recliner with his little lamp, on his laptop, just kind of scouring through stuff, taking notes and trying to figure it out." Finally, Prescott struck upon a disease that seemed to explain everything.

He took the information he'd found to his doctors, and they agreed to test him right away. The results came back and vindicated his hunch. Michael Prescott was indeed experiencing accelerated aging. He had a real disease with a real name.

■ ■ ■

At the turn of the twentieth century, while the folks of Punta Gorda were hawking the rejuvenating powers of their local fountain, a medical student named Otto Werner living half a world away published a medical report about a family who seemed to grow old at an accelerated pace. Four of the siblings, aged thirty-one to forty years old, had cataracts and gray hair. Two of them also had short stature. A few decades later, other physicians noticed similar cases, and the enigmatic ailment was given a name. It is now called Werner syndrome.

When Michael Prescott was finally confirmed to have this disease, it helped relieve the burden of not knowing what ailed him, but it also meant that he faced a terminal prognosis. In Werner syndrome, a person seems to age at fast-forward speed. They typically stop growing at puberty and see signs of hair loss by age twenty-five. At the same time, they also begin losing the fat under their skin and their muscles start to atrophy—all contributing to premature outward features of aging. By their thirties, many patients develop other early hallmarks of aging such as hardened blood vessels. People with this condition live, on average, until their early fifties.

Prescott forged ahead despite the worrisome prognosis of his illness. He continued cheering on his favorite football team and guided his son through life, reminding him often to approach the world with the right attitude. Ultimately, though, Prescott developed cancer—another common problem seen with Werner syndrome. Despite his perseverance, the medical complications of his ailments caught up with him. He passed away at age fifty-two, weighing only around sixty-five pounds at the time.

Although he hadn't known it for the majority of his life, Prescott was *born* with Werner syndrome: The root cause of the disease has been traced to inherited mutations in a single gene that makes a protein with important cell functions. The multitasking protein, called WRN, is thought to help ensure stability of DNA throughout the genome and may facilitate its repair as well. Werner syndrome mutations often result in a truncated, nonworking version of the protein. Without a functioning form of the protein, cells rapidly accumulate mutations. In other words, a person with Werner syndrome is born with a DNA error that lets *other* errors run wild in the genome. This error catastrophe is likely why the resulting complications of the disease are so numerous.

One school of thought is that DNA faults accumulate in cells over time, leading them to malfunction and contribute to the hallmarks of growing old. This school of thought has a major caveat: Not every disease that predisposes people to higher-than-normal mutation rates is associated with premature aging. The truth is that aging is probably a multifactorial phenomenon, rather than one caused by a single cellular mechanism.

That said, other examples linking mutation and aging do exist. Some compelling evidence supporting this connection can be found in the long-term health records of childhood cancer survivors. When they are young, many of them receive chemotherapy and radiation as part of their treatment, both of which are known to damage DNA. Later, as adults—when their cancer is in the far past—these survivors face an increased risk of premature frailty and low muscle mass, along with early signs of memory impairment.

Intensive cancer treatment and Werner syndrome are extreme examples in which mutations can accrue rapidly. But studying the bodily decline they can cause has pointed to a possible universal mechanism underlying aging. It suggests that mutations might be a culprit behind the physical deterioration seen in *all* people as they head into their later years.

Remember, you don't have to be sick to have DNA errors accumulate in your body. It happens naturally within every one of us. The big question is whether this causes the physical problems associated with becoming old. Throughout this book, we've seen how scientists have linked mutations to specific diseases, but what if mutation could also cause the widespread deterioration of multiple tissues and the generalized frailty seen in aging?

. . .

When Jan Vijg speaks, his sentences have the punctuated arc of upbeat arpeggios. As I listened to him talk at his desk in New York, at the Albert Einstein College of Medicine, in front of countless research journals stacked in piles, it was hard not to absorb a bit of his bountiful energy by osmosis. Vijg, who has published several hundred scientific articles, is part of a small but important lineage of scientists who have tried in recent decades to understand how mutation might be linked to aging. In his mind, a pivotal moment for the field came right after the structure of DNA was published. In the wake of that discovery, two famous physicists started to theorize about the possibility that errors in the genome might contribute to our senescence.

The first to do so was Gioacchino Failla, who was one of the first medical physicists in the United States and ran studies on the biological effects of radiation. During World War II, Failla was seconded to the top-secret government-run operation known as the Manhattan Project, which is well known now for developing the first atomic bombs. These weapons of war caused massive, horrific, and immediate death when deployed by the United States at the end of World War II, and have been linked to cancers that emerged afterward in those exposed to the bombs' mutation-inducing radiation. For some of the scientists within the Manhattan Project, including Failla, understanding the effects of mutations would remain an obsession.

When Failla reached his sixties, he began publishing papers on the subject of aging. He was concerned about the total radiation exposure that individuals experienced during their lifetimes and whether it could cause irreversible harm to chromosomes. These worries came through in his 1958 paper, which detailed his thoughts about how "hits" to the genome could perhaps explain the universal process of aging. The next year, *another* physicist who had worked on the Manhattan Project published a paper with a similar notion. The article was penned by Leo Szilard, a formidable thinker. (It was Szilard who had theorized about the nuclear chain reaction and patented the idea in the mid-1930s. He was also the person who had reached out to Albert Einstein and encouraged him to alert President Franklin Roosevelt that the Germans might be working on an atomic bomb, which eventually led to the creation of the Manhattan Project.) Like Failla's in 1958, Szilard's 1959 paper titled "On the Nature of the Aging Process" chalked up the cause of aging to genetic mutations.

Vijg has huge admiration for these giants of his field. But he notes that they ran into a crucial challenge back in the 1950s: "Szilard and Failla could not really test their hypothesis," he explains.

For a long while, the theories linking mutation to aging were just that—theories. They lacked hard evidence. By the early 1960s, that began to change. A major breakthrough came from an experiment by Howard Curtis and Katharine Stevenson at the Brookhaven National Laboratory on Long Island in New York. They looked at the liver tissue from mice and found that around 20 percent of the cells taken from mice older than a year had abnormal chromosomes, compared with only 10 percent of cells from young, five-month-old mice. In a follow-up experiment a couple of years later, Curtis and another member of the Brookhaven lab, Cathryn Crowley, found evidence that a short-lived mouse strain racked up chromosomal aberrations more quickly than long-lived mouse strains. These mouse studies provided some of the first evidence that genetic anomalies accrue as an organism grows old.

All of this was immensely intriguing. Yet despite the evidence from the Brookhaven labs that chromosomal irregularities increased with age, no one had figured out a way to test whether smaller DNA sequence changes were associated with growing old. Those tiny genetic sequence deviations in nonreproductive cells—also known as somatic cells—didn't really interest many of Vijg's contemporaries. "They all thought the frequency of somatic mutations is extremely low and could never cause aging," he says. He wanted to explore the role of these tiny changes but lacked the tools to do so.

Instead, Vijg spent the early decades of his career studying other genetic errors that accumulate with age. He coauthored papers on everything from DNA breaks in the brains of people with cognitive decline to UV-related damage in skin cells from old rats. This research suggested a connection between DNA damage and aging, yet it still didn't show small sequence changes accumulating over time.

Finally, with the advent of new genetic sequencing and engineering tools in the late 1980s, Vijg and his collaborators saw an in. Their approach relied on using a bacterial gene called *lacZ*. The scientists designed mice so that their DNA contained this gene, which would act as a reporter for how much mutation accumulated in the rodents' genomes over time.

The scientists took cells from the genetically engineered mice at different stages of life—including those that had reached almost three years old, a ripe old age for a mouse. Then, through a series of steps, they sequenced the *lacZ* gene the rodents carried. The results indicated that the animals' liver cells accumulated genetic changes throughout their lifespan. The researchers also saw a rapid increase in a particular kind of genetic reshuffling known as rearrangements at twenty-seven months (toward the end of the animals' lives).

The experiment represented the first time that scientists were able to show a quantifiable increase in tiny sequence mutations in different organs and tissues of animals during aging. Follow-up data from subsequent studies in mice run by Vijg and others would show that this increase was happening in more than just the liver. Mutations were found to accumulate with age in virtually *all tissues*.

During his long career, Vijg has spent countless hours pondering *how* mutations might result in the breakdown of the body over time. One of his main theories is that among the errors that accrue with age, those affecting the parts of our DNA that normally regulate the function of genes have enormous consequences. He explains that while only 1 percent of the genome encodes proteins, around 10 percent is devoted to these regulatory sequences. As more and more mutations hit those regulatory regions over time, genetic activity within cells may go haywire.

Vijg says this theory about age-acquired mutations creating "noise" in the cell is a throwback of sorts. It's partly informed by the work of the British-born biochemist Leslie Orgel in the 1960s. Orgel proposed what is now known as the error catastrophe theory of aging. According to this theory, cells create faulty proteins by mistake that then cause a feedback loop creating even more inaccurate proteins, ultimately causing an organism to die. Essentially, his idea was that cell function goes off the rails as we age. Although Orgel's hypothesis has never been put to the test, it's echoed in Vijg's speculation that accumulating mutations cause a spiraling malfunction of our genetic activity.

According to Vijg, there's another source of genetic noise—in addition to the mutations within our DNA code—that may add up over time. He says that a contributor to this noise might be alterations to the chemical groups that sit *upon* our DNA, known as epigenetic marks, which can be influenced by our environment and behavior. Interestingly, researchers have found patterns of epigenetic change associated with aging. (These changes have garnered so much interest as predictors of age that some scientists refer to them as "epigenetic clocks.") Moreover, in one analysis of data from more than nine thousand people, the hot spots of mutation within the genome corresponded to the areas with epigenetic marks linked to aging. It's unclear whether changes to the genome precipitate epigenetic changes or vice versa, or whether these are both downstream effects of an earlier event. Whatever the source of the genetic noise might be, the hope is that untangling the various ways time leaves its mark on the genome will help us better understand how to potentially prevent or reverse aging.

The quest to show how the mutations we accumulate might cause us to grow old—and what might be done about it—has been picked up by a whole new generation of scientists. And they are looking not just at humans and mice. They're testing far-flung branches of the animal kingdom.

■ ■ ■

Back in the late 1950s, when Gioacchino Failla wrote about his theories of aging, he believed that mutations happened more rapidly in short-lived species like rodents compared with long-lived ones like humans. Six decades later, scientists found a way to investigate this possibility on a grand scale.

One of them was Alex Cagan. For two years, Cagan waited with bated breath for express mail packages containing tiny samples of animal intestines packed on dry ice. The deliveries arrived, sometimes many weeks apart, but each brought him closer to his goal. By the autumn of 2021, the laboratory where he worked at the Wellcome Sanger Institute, half an hour south of Cambridge, UK, had accumulated a veritable Noah's ark of preserved intestinal specimens. Giraffe guts? He had some of that. Samples from lions, lemurs, and tigers? Those had come, too. The animals had died from natural causes or illnesses of aging, some delivered from the London Zoo. Cat, cow, mouse, and dog intestine had also been received, along with many others, including that of a horse and a harbor porpoise. There were even intestines from primates like the black-and-white colobus monkey, which has fringes of hair that hang on its back reminiscent of an Elvis jumpsuit. As soon as each package arrived, Cagan and his collaborators would open the inner white Styrofoam casing—releasing a small cloud of fog coming off the dry ice—then take the piece of pink gut tissue, about the size of a pencil eraser, out of the test tube and get to work. But they were not interested in studying digestion. They believed these guts held a secret about aging.

Cagan is a geneticist, albeit one with an artistic flare. He has a penchant for brightly colored Patagonia fleece pullovers and wears a watch with an orange wristband. On weekends, he attends a class where students sketch a live model, and his illustrations of scientific concepts have been featured on the covers of famous research journals. Cagan got to put his drawing skills to use in the lab when annotating the gut tissue to mark cells for analysis. After his teammates would prepare ultra-thin slices of the intestinal tissue onto microscope slides, Cagan would put them in a machine with a display screen where he could superimpose lines he sketched on a connected tablet with a digital stylus. The slides lacked a bottom cover, so Cagan could draw around a specific area where he wanted to sample, and a laser beam would knock the circled cells—about one hundred to one thousand at a time—down into tiny wells in a laboratory dish positioned underneath. Cagan's particular obsession was with zeroing in on tiny pockets, known as crypts, situated all over the inner lining of the gut. He would trace around those U-shaped structures on his display screen and then—*zap!*—use the laser to drop the cells from the underside of the slide into the lab dish below. From there the cells would get whizzed along for DNA analysis.

Despite their spooky name, intestinal crypts have real scientific appeal for geneticists: Each of these microscopic structures grows from a single stem cell, making it possible to see which shared mutations have arisen in those cells. Cagan and his teammates, including his boss at the time, Iñigo Martincorena, collaborated to sequence the DNA of about four crypts per each animal, averaging around a dozen crypts per species. The team then sequenced cells from muscle tissue in the gut lining, to help them compare and see how many mutations were occurring in the crypt cells. They also uploaded data from intestinal crypts from twenty-eight deceased human donors. When they crunched all the data, they saw a surprising trend: Across all species, the animals who reached old age accumulated about 3,200 mutations within each crypt during their lifespan. The similarity of that total at the finish line was shocking.

The data also helped explain *why* the total number of mutations across species was so similar: As Failla had once predicted, shorter-lived species accumulate mutations at a faster *rate*. Mice appear to acquire around eight hundred mutations in the stem cells of their intestinal crypts each year of their brief life. Dogs—which are considered to be very old if they reach around fifteen years—accumulate about 250 such mutations annually, whereas giraffes, which live to around twenty-four years, amass approximately one hundred during the same time. This process unfolds much more slowly in long-lived species. Humans, for example, average just forty-seven mutations each year in a given intestinal stem cell crypt. Cagan was stunned to see the data line up so neatly. "I was expecting that there would be *a* relationship with lifespan, but that it wouldn't be as strong as what we found," he says.

Some animals mutate faster and some slower, but the inverse relationship between lifespan and mutation rate means that they reach their ultimate end with about the same total number of these genetic changes. It's almost as though Mother Nature has an equation in her back pocket to keep things even.

The comparative analysis of different species by Cagan and his teammates has strengthened the theory that the pileup of DNA errors over time makes us grow old. But he and his coauthors believe it could be a bit more complex. They speculate that the aging of an individual might not simply be caused by rising numbers of mutations. It might be due to the accumulation of specific mutations that cause affected cells to become damaging and selfish. By this theory, the mutated cells greedily replicate and take over tissues, ultimately outcompeting surrounding healthy cells. Others, including Jan Vijg, point to this possibility. And the evidence to support it is growing.

■ ■ ■

One of the illnesses that most of us think about as we become older is cancer. Advanced age is *the* most prominent risk factor for cancer overall. The incidence rate of the disease climbs throughout life. For people in their late forties, it is around 35 per 10,000. But that shoots up to around 250 per 10,000 in those in their early eighties. As we head into our golden years, we unfortunately have to fend against cells that have picked up nasty mutations and multiplied into malignancies in our tissues. We go for our annual mammograms or prostate exams, and for colonoscopies to check for suspicious polyps. None of it is glamorous, but it is necessary.

It's not just cancer, though. In our modern era of genetic sequencing, we've learned that there's a whole range of illnesses that transpire below the surface of our skin when cells go rogue. You may recall how in an earlier chapter we encountered the aging-associated phenomenon called clonal hematopoiesis, in which mutant cells gradually take over the body's blood supply. Clonal hematopoiesis is far more common in people who are in their retirement years. It has been linked to cancer, but also to other ailments that commonly affect older people, such as heart failure and the buildup of plaque inside blood vessels.

This phenomenon might go beyond the blood. It might contribute to islands of mutated cells within the brain. An analysis of postmortem brains published in 2022 found an intriguingly similar pattern of mutation in that organ with aging. Sixteen percent of the brains from people sixty or older were what the researchers called "hypermutated"—essentially having many cells with genetic mistakes. Meanwhile, only 2 percent of the brains from those under forty met that definition. The scientists behind the study believe these hypermutated brains occur in people within whom neurons with certain DNA changes have begun to outcompete others, just as mutant blood cells do in clonal hematopoiesis.

More recently, a team including the bioinformatician August Yue Huang and Christopher Walsh at Boston Children's Hospital helped connect some interesting dots. The researchers uncovered recurrent mutations in genes implicated in clonal hematopoiesis in cells from the prefrontal cortex brain region of patients with Alzheimer's disease. The study was particularly damning of a subtype of cells known as microglia. Microglia typically function as the immunity heroes of our nervous system. They constantly patrol for pathogens and damage. But if something goes awry in their internal programming, they can get overaggressive and cause inflammation. Having too many of them might also be a problem.

Huang, Walsh, and their collaborators noticed that up to 40 percent of the microglia from Alzheimer's patients had been hit by mutations in cancer-associated genes. Some of these genes were the same as those affected in clonal hematopoiesis. This overlap suggested that the genetic changes in the microglia might enable them to replicate in a clonal way, making many copies of themselves and outcompeting their normal counterparts. The researchers call the phenomenon mutation-driven microglial clonal expansion—quite a mouthful—or by its acronym, MiCE.

The finding of mutations within the patients' microglia fits with the results of Walsh's past research, which linked the accumulation of certain DNA changes in neurons to a heightened risk of cognitive ailments. "We think that this is a key new way of looking at aging and common forms of neurodegeneration like Alzheimer's disease," Walsh says. Past studies from him and his collaborators have suggested that neurons from people with Alzheimer's rack up at least a couple hundred more mutations than expected in people who remain cognitively healthy—about the equivalent of more than a decade of excess DNA error accumulation.

It can feel overwhelming to contemplate all the messiness that arises in our chromosomes with time. But we are not totally defenseless against mutations in our genomes. There's a counteracting force to some of the DNA damage that we experience. Each of us possesses machinery in our cells that can mend our genetic material, and we're learning more each day about how it can restore the integrity of our genes.

. . .

Although Earth formed around 4.5 billion years ago, it had to wait a long time before it got its sunglasses. The ozone layer—the part of the stratosphere that shields the globe from most of the sun's ultraviolet radiation—became substantial only around five hundred million years ago. That's long enough ago for our species, but the first organisms on the planet had to tough it out in a harsh environment lacking that protective barrier. Our single-celled ancestors had to develop mechanisms to cope with the multiple ways that ultraviolet radiation can screw with DNA. That became even more necessary as they evolved more complex cellular functions, requiring longer genomes to encode for the more sophisticated inner workings. Longer genomes meant more places for DNA damage to occur.

The solution to such damage was DNA repair. It evolved as one of the earliest genetic traits, according to Jan Vijg. This kind of repair machinery became even more necessary after the emergence of multicellular organisms. More cells meant more places for damage to happen, including in nonreproductive cells.

Ultraviolet radiation, which early organisms had to cope with—and which, to a slightly lesser degree, still bombards us today—is nasty. It messes with DNA in multiple ways. Sometimes it causes molecules of the sequence to crosslink with unusual partners, distorting the structure of the strands and introducing bends and kinks. Other times it encourages the production of damaging free radical molecules that can throw off DNA replication and even lead genetic sequences to break.

Thanks to DNA repair machinery—the hero inside every cell—the genetic aberrations caused by ultraviolet radiation can often be overcome. Cells can recognize DNA damage that involves breaks, losses of DNA molecules, or crosslinks between bits that shouldn't occur. By some estimates, each cell in the body encounters one hundred thousand of these kinds of errors each day. "Most are repaired within minutes and certainly hours," according to Vijg.

DNA repair can involve an orchestration of dozens of molecules to get the job done. Take, for example, the thirty or so proteins that work to perform what's known as nucleotide excision repair. This process spots bulky DNA alterations, clips out the corrupted fragments, fills in the necessary replacement sequences, and seals them in place.

Not all DNA aberrations are equally possible to rectify. When small mistakes are made at single points in the genome—for example, the wrong DNA molecular subunit gets substituted in the long strand— the cell doesn't have a good way to know that an error has crept in. In these instances, and even in cases where there are large chromosomal aberrations, the original template for the correct sequence is lost forever. The right sequence can never be restored.

We know that our genetic mending machinery is important, because people born with faulty versions of it have devastating and life-shortening conditions. Problems with DNA repair are found in Werner syndrome, the disease that affected Michael Prescott.

Cockayne syndrome is another heartbreaking example, and it manifests even earlier. In the classical form of the disease, symptoms first appear after an infant turns one year old. Suddenly the child seems to start aging rapidly, but without growing into an adult. Those with

Cockayne syndrome ultimately can have sensitivity to light, dwarfism, and signs of premature aging such as hair that turns gray well before adulthood. The life expectancy for this form of the disease is only ten to twenty years, but individuals still develop complications typically seen in people decades older, sometimes including cataracts in their eyes (reminiscent of Werner syndrome). In the 1990s, researchers successfully pinpointed that children with this condition are born with broken DNA repair genes.

Inheriting faulty DNA repair machinery seems to be a recurring phenomenon in diseases of accelerated aging. The opposite might be true in cases of impressive longevity: It's possible that some variants of DNA repair genes might *enhance* one's ability to live to a very ripe old age.

Could DNA repair molecules that counteract mutation hold the key to longevity? It's a very speculative notion, but there are some hints it could be true. In one 2021 study, scientists sequenced the whole genome of more than eighty centenarians from Italy. The average age of the volunteers in the study was above 106 years old. It turns out that the centenarians were more likely than ordinary folks to carry certain—potentially beneficial—versions of a gene involved in DNA repair. Previous research has suggested that the enzyme made by the gene is activated in response to a huge, aforementioned source of mutation: ultraviolet radiation.

It's a tantalizing finding. Perhaps if we could simply safeguard the integrity of our DNA, we might be able to add years to our lives. In reality, the evidence for this idea traces back decades. Soon after scientists like Failla and Szilard started talking about the accumulation of mutations occurring with age, researchers began to theorize that DNA repair could perhaps extend life. Some, like Richard Setlow, a biophysicist who worked at the Oak Ridge National Laboratory in Tennessee, felt that there was a direct correlation between the lifespan of a species and how efficiently it could repair damage to its DNA.

As you may have noticed by now, scientists love to compare animals when it comes to longevity. Setlow and his collaborator Ronald Hart decided to look at the capacity for DNA repair in skin cells from seven different species. These included a mouse, shrew, rat, hamster, cow, elephant, and a ninety-five-year-old man. The pair exposed the skin cells to radiation in the lab and measured a proxy indicating when genetic fixes had been made. Setlow and Hart found that, among the species they studied, an increased capacity for DNA repair correlated with a longer lifespan. They published the results in 1974. However, they were cautious not to overinterpret the findings. "The connection between DNA repair and aging is only that, a connection," Setlow said in an interview with *The New York Times* more than a decade later. "It's not clear whether long-lived species are that because of DNA repair, or that they have good DNA repair because they live a long time."

Today, the search for an answer is very alive. And some of the newest inklings of the role of DNA repair come from studying how it operates in the longest-lived mammal species on planet Earth.

■ ■ ■

The lab where biologist Vera Gorbunova runs experiments houses some exceptional animals. Inside the space where she and her husband Andrei Seluanov work together at the University of Rochester in upstate New York, you'll find naked mole rats from East Africa. These rodents have two protruding upper teeth jutting out of their mouths, as well as two lower incisors, which they can move independently, like chopsticks. They are nearly hairless, and their pinkish skin sags and wrinkles around places like their necks and hind legs. But even if the naked mole rats are not going to win any beauty contests, they are champions of longevity. Naked mole rats have the longest lifespan of any rodent species and can live beyond thirty years.

Gorbunova has devoted her career to investigating longevity, including how genes that guard the integrity of DNA might extend the lifespan of a species. Her work has taken her around the globe, including to some extremely cold places. Gorbunova has a strong dislike of the cold, despite having grown up in the city now known as Saint Petersburg, Russia, and working as an adult in snowy Rochester. "I absolutely prefer to be in the tropics," she confessed to me. In spite of her aversion to frigid temperatures, she packed her bags in 2014 for a voyage far north.

Gorbunova was headed to Utqiagvik, Alaska, home to the largest community of Indigenous Inupiat people. Its location—on the coast of the Chukchi Sea—is nine miles from the northernmost point in the United States and more than three hundred miles north of the Arctic Circle. Getting there was no cakewalk, she recalls. "You fly to Anchorage, which is a pretty typical flight, but then you have to get on the plane that goes north," she explains. "And this plane, half of it is passengers and half of it is cargo, because there are no roads leading there. So all the supplies, all the groceries have to be flown in."

Gorbunova describes her arrival there as memorable. "The airport is amazing," she says. "Sometimes they get polar bears walking on the runways." The scenery moved her, despite her dislike of the cold. "It's all tundra and permafrost. And there are no trees," she adds. "Oh, it's beautiful—it's *beautiful*—but it's very different from what we're used to."

She had traveled there because the sea waters around Utqiagvik are inhabited by one of the most remarkable species on the planet: the bowhead whale. These animals, which can weigh more than 80,000 kilograms (almost 180,000 pounds) are the second-heaviest creatures on Earth, right after the blue whale. And while the bowhead whale might come in second for mass, it earns first place for longevity: It's the mammal with the longest lifespan on Earth. One bowhead whale reached 211 years old, and some genetic clues suggest whales of this species could have a maximum lifespan of 268 years. Gorbunova and her colleagues wanted to get hold of cell samples from the animals for this reason. "We were very interested in bowhead whales because of their longevity and their size," she says. "This is the only mammal proven to live longer than humans."

After arriving in Utqiagvik, Gorbunova met with local Indigenous leaders about her potential project. The community is one of the few that can legally hunt bowhead whales, and they do so using a mix of modern and traditional methods, navigating the waters in very slight boats. They kill only around several dozen a year. Gorbunova marveled at the remnants of the animals along the coast. "There are bones there, and they're majestic," she says. During her visit, she walked between the ribs of the whales on the beach and sat inside the ossified remains of their mouths. An agreement was made with the local community that her team could return to receive small samples from the whales that were harvested in the future for subsistence.

When Gorbunova's students made subsequent visits to Utqiagvik, they collected the samples from the lungs and skin of the bowhead whales as soon as the carcasses were brought ashore. They took only a few grams of each. That was all they needed. From there, the samples remained in a container kept at 4 degrees Celsius—cool enough to preserve the cells but warm enough to keep them alive during the transport across the continent on their way to Rochester.

Once Gorbunova and her team returned, they coaxed the whale cells to replicate in the lab. The scientists observed the cells and remarked on their impressive features. First, the bowhead cells mend breaks and mismatches in their genetic sequence extremely well. Second, the bowhead cells contained high levels of two proteins that increase the efficiency and accuracy of DNA repair. When the researchers exposed human cells to these proteins, the same benefits appeared. The whale cells simply had a naturally big stock of these proteins. "One in particular was so much more abundant in the whale," Gorbunova explains. "There was about a hundred times more of it" than in human cells, she adds. The exceptional protein is known as cold-inducible RNA binding protein, or CIRBP for short.

CIRBP helps cells respond to the stress of cold temperatures, so Gorbunova reasons that bowhead whales have a knack for churning it out because they live in icy waters. When human cells are exposed to cold temperatures, they also kick up their production of CIRBP. (Gorbunova acknowledges that some people swear by the health benefits of dunking themselves in cold water, but it is far too premature to say whether the effects of that practice are linked to CIRBP.)

The group posted their results online in 2023. One of the coauthors and collaborators on the project was none other than Jan Vijg. That year, Gorbunova and Vijg joined other contributors to conduct a major review of ways to mitigate the age-related burden of mutations. "The most logical approach to improve the maintenance of genome sequence integrity is to seek ways to upgrade DNA repair," the group wrote. "Exceptionally long-lived species could tell us how to reduce mutation burden by upgrading key nodes or master regulators of genome maintenance."

The thought hasn't escaped Gorbunova that proteins such as

CIRBP—if they do indeed counteract DNA damage—could perhaps have a place in modern medicine. But CIRBP isn't the only cellular protein that fights back against mutations, and Gorbunova isn't the only researcher who sees pharmacological potential in DNA repair molecules. As scientists pull back the curtain and see how damage to the genome increases with age, more of them have felt tempted to find ways of combating the accrual of those genetic errors.

There seems to be the beginnings of a gold rush to try to fix corrupted DNA in our bodies. A handful of companies have launched with this ideal in mind. "For those of us who are trying to turn back the clock and slow the aging process down, DNA damage can be one of the main things we fight against," says the website of one such start-up, called Genflow Biosciences. "But lucky for us, we live in an age where innovative biotechnology has made DNA damage repair possible." It's an exciting proposition. But as we will see, there might be a tough trade-off for this kind of power.

■ ■ ■

For as long as humans have searched for the Fountain of Youth, people have tried whatever they believe it takes to avert the inevitability of aging. In modern times, seekers of vitality are unlikely to look for a mystical place with restorative waters. Instead, they peer at the tiny molecular machinery of cells to hunt for a solution. That includes some scientists who believe that activating DNA repair might reduce wear and tear on the genome and therefore perhaps extend life, too. This is the antiaging philosophy behind Genflow Biosciences, which launched in the United Kingdom in 2020. "By treating ageing as a risk factor to disease," the start-up says on its website, "Genflow Biosciences aims to reduce the financial, emotional, and social costs of an ageing population."

The company already has a few compounds in the development pipeline, one of which aims to treat Werner syndrome, the inherited genetic condition of accelerated aging that affected Michael Prescott. Genflow Biosciences is aiming to start clinical trials on a compound to reverse damage to the liver and another one it hopes will have anti-aging effects in dogs. Although it's exploring different applications of its approach, there's a common thread: All the drugs it has in development work through a cellular pathway regulated by the *SIRT6* gene, which makes the Sirtuin-6 protein that helps guide DNA repair.

One of the people who sits on Genflow's scientific advisory board and who has studied the role of *SIRT6* is Vera Gorbunova. In 2022, she was one of the senior leaders of a genetic sequencing project that found some centenarians possessed a rare variant of the *SIRT6* gene that enhances genomic stability. Jan Vijg was also among the resulting paper's coauthors. Genflow envisions extending life by giving people a gene therapy that delivers this version of *SIRT6* found in human centenarians.

Other researchers hoping to create antiaging therapies have turned to gene editing. One company has started designing a treatment that relies on the CRISPR/Cas9 genome-editing system. This technology has already been deployed in an approved treatment for the blood disorder known as sickle cell disease, but there's a hope it could be tailored as a genome-surveillance tool to look for—and correct—mutations. The venture, aptly named Spellcheck Bio, counts George Church, a geneticist at Harvard University and serial entrepreneur, as one of its cofounders.

Beyond DNA repair and gene editing there are even more ideas about how to reverse the genetic changes associated with aging. Scientists now talk about "cellular epigenetic rejuvenation." This refers to a theoretical approach in which the chemical markers on the genome are restored to a pattern found in cells of more youthful individuals. Potential downsides to this theoretical treatment exist. Epigenetic marks help guide the function of cells, so an indiscriminate reformatting of such marks might cause cells to lose their cellular identity and become malignant.

Then there's a proposed way of reducing age-associated genetic changes that would take a blunter approach: getting rid of cells that have accumulated too many mutations during a person's lifetime. Doing this would require finding drugs that possess sniper-like accuracy. Ideally, they would kill the highly mutated cells without harming healthy tissue.

For a long while, some scientists have pinned their antiaging hopes on antioxidants, which come from a wide range of sources that include fruits and vegetables. These molecules can clear away DNA-damaging free radicals, thereby preventing mutations. Researchers have tested a bountiful range of antioxidants—everything from blueberry extract to apple polyphenols to tangerine peel—in various species to see whether they can boost longevity.

Unfortunately, the evidence about whether antioxidants can prolong life is murky at best. In fact, in some experiments these compounds actually shortened it. And although a number of studies—especially those conducted on invertebrates such as fruit flies—have suggested that antioxidants might help increase average lifespan, there's relatively little data to support that antioxidants can extend the *maximum* lifespan of a species.

Despite the mixed results of studies of antioxidants, there's still loads of interest in their possible antiaging effects. One thing that sets apart antioxidants from DNA repair molecules that reverse genetic errors or methods that eliminate heavily mutated cells is that antioxidants have the potential to stop mutations from occurring in the first place. In this way, they might offer a tantalizing chance to prevent time—in the form of genetic changes—from leaving its mark on the genome.

That, increasingly, seems to be the lofty goal that antiaging proponents want to chase: to preserve the integrity of DNA against all odds. After all, experiments ranging from the mouse studies in the 1960s to more recent investigations of giraffes and bowhead whales have painted a picture in which mutation seems inextricably tied to growing old, and in which staving off these genetic changes could hold the key to longevity. Yet the simple idea of slowing or preventing mutations could come with all sorts of complications. In biology, everything has a cost. Trying to rein in change within a genome, made of billions of subunits that are prone to wear and tear over time, can seem almost as fantastical as finding a Fountain of Youth with magical healing waters. Still, some modern-day scientists, like explorers centuries before, are determined to succeed.

■ ■ ■

If there is one overarching message of this book, it is that our genomes are surprisingly dynamic. For almost a century and a half, since people like the pathologist David Paul von Hansemann first hypothesized that chromosomal instability was linked to cancer, we have been slowly awakening to the idea that the instructions of life within our cells are subject to change. The work of early twentieth-century cancer biologists revealed that genetic differences can exist among cells in the same organism, chipping away at the illusion that the instructions encoded in chromosomes are the same for every cell in the body.

Scientists in other fields of biology also found strands of evidence for genetic diversity within organisms. It's impossible to end this book without mentioning a key insight of the 1940s from the plant scientist Barbara McClintock. McClintock was passionate about understanding the patterns of genetic traits in corn. She was so invested that she slept in a sleeping bag by her research crops when the scarecrows failed to keep hungry raccoons away at night. McClintock meticulously tracked the different pigmentation patterns in multicolored maize (the kind that you might find as a festive ornament on doors during autumn holidays). Through her recordkeeping of the kernels' purple and brown hues, she developed a theory that genes she called "controlling elements" could move along chromosomes—and in doing so create new color patterns. The amount of spotting on a kernel would be influenced in part by how early the mutation happened during development. Those in which the spot or streak of color was bigger had experienced the mutation early on, so it affected more of their cells. Controlling elements would later become known as "jumping genes," and similar transposable elements are now known to operate in human cells, especially during early embryonic development. When they jump into the wrong places, they mess up the activity of important genes. Although McClintock's ideas were not initially well accepted by her peers, she eventually became the first woman to be the sole recipient of the Nobel Prize for Physiology or Medicine. It was further validation of the notion that DNA is far from a static molecule.

It's not just that DNA can change—it's that when it does, it can sometimes endow cells with an advantage within the body. These mutated cells can churn out clones of themselves, and we're increasingly aware of their influence on our health. When Peter Nowell put forth his theory about how tumor cells compete with one another (not only with healthy cells), he wrote about this resulting in a "clonal evolution" in which the most aggressive cancer cells win out. Beyond cancer researchers like Nowell, immunologists and hematologists in the 1970s further cemented the idea that cells with certain mutations could scale up their numbers with huge consequences. In the case of immunology, an idea emerged that when a cell acquired genetic alterations enabling it to make an effective antibody in the presence of an invading microbe, it would be promoted to make clones of itself, which would in turn produce even *more* of that antibody, thereby shoring up the body's defenses. With hematology, work from Lucio Luzzatto and others helped reveal that mutated blood cells can predominate a person's blood over time, leading to a life-threatening form of anemia.

All this discovery about how mutant cells take off in the body to produce clones took place in disparate medical fields—oncology, immunology, hematology, etc.—so it happened largely under the public's radar. But a lot has changed in the past decade, as technologies to sequence individual cells have become widely available and different disciplines have uncovered that specific mutations in clones appear in seemingly unrelated conditions. There's a convergence taking place. Take, for example, the blood condition CHIP, which we encountered in chapter 4. CHIP describes a condition in which clones of a certain type have reached a threshold level of a person's blood. Clones in CHIP that have a mutation in the gene *TET2* have been linked to an increased risk of heart failure. We also now know that *TET2* mutations appear in some clonal populations of cells in the brain, and there's interest in whether this might contribute to—or protect against—disorders like Alzheimer's disease.

There's another reason all of this is becoming more talked about: It turns out that conditions involving acquired mutations are much more common than previously thought. For example, as many as one in five people over the age of seventy might have detectable CHIP mutations. Some researchers go as far as to suggest that everyone over the age of forty might carry stem cells with these kinds of mutations. The knowledge that there is a creeping accumulation of mutation—and that it lands in unpredictable locations within our genomes—might feel disconcerting at times. But it's a reality we cannot ignore. This truth is more clear today than ever because of the mounting evidence of how acquired mutations can influence health outcomes.

■ ■ ■

Most of this book has focused on how the mutations we pick up, either in the womb after conception or as we grow and age, can reverberate within the body and change our *physical* well-being. But, as neuroscience has illuminated, noninherited genetic changes have the power to shape our *cognitive* health as well.

The disruptive effects of acquired brain mutations are gradually coming into focus. Studies are showing that the head is not some haven where cells remain pure and untouched by DNA damage or by mistakes that get introduced during sequence repair. There's a mess of mutations that can accumulate up there, just like other places in our body. Sometimes mutations strike early in embryonic development, leading to problems such as epilepsy or autism. Other times they may accrue in life and perhaps contribute to conditions such as Parkinson's.

It's a bit uncomfortable to think of the brain as such a genetically dynamic organ, and of what this means for the mind. Of all our different body parts, our brain is the one that is most intimately tied to our sense of self. Our worldview is shaped by the collection of cells inside our head working in concert to make sense of our surroundings. The very essence of our personality and our identity is encoded by the brain's neurons. Maybe for this reason, it's somewhat jarring to learn that the cells that make up the organ might not all be the same—and that they differ genetically more and more over our lives. In a way, it suggests a fractured sense of self. Each of our brains contains a multitude of genetically divergent cells working together to form who we are.

When I titled this book *Beyond Inheritance*, I wanted to emphasize that our genetic destinies are not necessarily defined by what we inherit from our biological parents. The more we uncover about the genetic errors that arise within us as we grow and age, the more we understand that these mutations have the power to shape our individual health trajectories. And the outcome of mutations is not always bad. Their ability to right a wrong has gone largely underappreciated. As we learned in chapter 6, sometimes randomly acquired genetic changes can resurrect a defunct gene. This almost miraculous event can undo a life-threatening immune disorder or completely eradicate a dangerous form of anemia.

There's a tremendous amount of momentum among medical researchers in recognition of the importance of somatic mutations. One big moment came a few years ago, when the US National Institutes of Health launched a major initiative to study this biological phenomenon. The program is known as Somatic Mosaicism Across Human Tissues, or by its acronym SMaHT (which sounds a lot like "smart" said with a Boston accent). The government allocated $140 million for the five-year program to help make sense of the rapidly growing data on somatic mutations and how they affect us. Perhaps no one captured its timeliness better than biologist Kenneth Walsh of the University of Virginia. When quoted about the launch of the initiative, he pithily summed up our evolving understanding of how our genetic makeup changes over a lifetime. "The genome you are conceived with," Walsh said, "is very different from the genome you die with."

* * *

In the last century, scientists were desperate for ways to document any somatic genetic changes. Now, with the advent of modern DNA sequencing technologies, an opposite challenge has emerged: They are desperate for ways to make sense of a deluge of data. In just the last few years, the time it takes to get readouts of the somatic mutations in a tissue sample has shrunk from several days to mere hours. It's an embarrassment of riches.

Suddenly, researchers have more information about the myriad genetic anomalies in each person than they ever dreamed possible. Finding a signal in that noise—in other words, figuring out which mutations are meaningful for a person's health and which ones make no difference at all—is perhaps the single biggest challenge this field faces going forward. We're quickly going from not appreciating the genetic diversity in the body to struggling to sort through the clutter now that we see how much of it is going on.

One person determined to help with the flood of this information is Shixiang Sun, a member of Jan Vijg's current laboratory nestled in the middle of the Bronx. Sun's quiet, spartan office is off a hallway from where other people in the team are busy pipetting samples into machines such as the CellRaft AIR System, a device the size of a large suitcase that allows them to select single cells for genetic sequencing.

While other researchers on the team generate data, Sun works on a project to make sense of it—as well as data on somatic mutations submitted by other labs around the world. The effort is known as SomaMutDB (the DB stands for database). Sun and others have grown it to include more than nine million somatic DNA variants cataloged from twenty normal human tissue and cell types.

The SomaMutDB catalogs the mutations by aligning them to their respective positions along the three billion spots of the human genome. For example, it shows that a study has found a mutation in which a DNA subunit called guanine (G) is changed to one called adenine (A) at position 85,249,536 on chromosome nine. By cataloging somatic variant data from numerous research projects around the globe, Soma-MutDB is starting to make order of the chaos.

Ironically, this monument to the genetic variation that exists within people and their different tissues wouldn't be possible without the work of the Human Genome Project, which once stood as a symbol for the common code within all humans. The benchmark onto which Soma-MutDB maps variants is the same reference genome that scientists have been fine-tuning since they first announced its initial draft sequence a quarter century ago. At the time it was first published, the reference genome was celebrated as a shared sequence across humankind, but now that we have the technology to document mutations within our tissues, it's become a tool to navigate the genetic diversity among—and *within*—individuals.

It can be hard to show the influence of single DNA changes, so SomaMutDB hopes to facilitate the discovery of mutational "signatures." These signatures are essentially trends of genetic change spread across the genome that can be associated with, for example, exposure to dangerous substances, or with illness itself. Essentially, they're patterns you can see once you sum up all the separate alterations.

Previously, research groups have shown mutational signatures linked to cancer risk factors. For example, scientists from the Wellcome Sanger Institute have documented how certain forms of ultraviolet light exposure seem to shift cytosine (C) to thymine (T), and how tobacco smoke exposure is linked to the replacement of cytosine (C) in the genome with adenine (A). Moreover, the fingerprint of change seen with tobacco smoke is different from that seen with chewing tobacco. The hope is that if scientists characterize the types of DNA errors caused by mutagens like these, they might be able to develop better ways to protect against that damage.

In the future, we might see more of this kind of thing: research that goes beyond single mutations to instead make meaning of the constellations of small genetic changes that accumulate in tandem across the genome. There is a whole universe of DNA to explore.

. . .

You might think that scientists would see the ever-accumulating mutations within the body as too daunting to rein in. But humans are often inclined to control natural processes—or at least to attempt to do so.

Throughout the book we have met researchers who are not only studying mutations happening in the body but are also trying to shift them. Sometimes they aim to prod cells toward good mutations. For example, we encountered immunologists designing vaccine boosters to nudge cells' genetic code toward producing better antibodies. Other experts want to repair errors in DNA. As we saw in antiaging efforts, a small group of investigators are starting to see whether they can harness the genetic repair machinery in cells to somehow extend life. In yet other fields, the objective is to stave off nefarious genetic changes in cells. We met oncologists and radiologists seeking to do so by adding careful pauses in their patients' cancer therapy. Those breaks are meant to remove evolutionary pressure on tumors, and thereby slow down their tendency to acquire mutations that make them impervious to treatment.

As you can see, different visions of influencing mutation have emerged within medicine. Because of this, the hero of genetic change in one discipline can be the villain in another. One case in point involves the enzyme called activation-induced cytidine deaminase, or AID for short. You might remember it from chapter 3. AID encourages a process known as somatic hypermutation, which helps immune cells switch up their genetic sequence to make new and better antibodies against microbes. Clearly AID has a useful role. But in stimulating mutation, AID might also encourage the development of certain cancers. So some researchers have attempted to find a drug that can inhibit it. In one experiment, they tested more than ninety thousand compounds and identified candidates that could mitigate AID activity. It's all very early days for this kind of exploration, but it exemplifies how various groups have decided to tackle sources of genetic change that other groups see as beneficial.

After billions of years of life on Earth, humans are the first living creatures that seek to shape their genetic destinies. We already achieve this by modifying our behaviors, like quitting smoking or slathering on sunscreen. However, we may need to ask whether it would always be wise to block or erase mutations with the possible gene-editing or drug interventions being contemplated. And we need to ask if we could do so with enough precision. Genetic change is necessary for a functioning immune system and has a role in restorative processes such as liver regeneration. Can we be specific enough to eliminate just the bad mutations while keeping the good? If we decide to choose which ones to block or erase and which ones to let run wild, then we will have to do so with care. This is, of course, assuming that we will be able to develop the technology even to make that choice.

We also have to contemplate the metaphysical implications of our potential mutation-blocking powers. In the past, people have referred to scientists using gene-*editing* technology—which involves adding or subtracting code from the genome—as "playing God." But it's equally remarkable to *prevent* the genome from changing. If we were to use new medicines to reliably and safely stop mutations in our DNA, would that not also be a godlike intervention?

There is another philosophical matter to grapple with here as well. Mutation is what helps species evolve. We cannot forget that reining in mutation in reproductive cells would theoretically slow the evolution of new traits. It's probably not a concern for those developing medical treatments that protect or repair patients' DNA. Humans live in relatively stable environments and, as such, may not face the same pressure to rapidly adapt as other species. But we shouldn't lose sight of this potential consequence of mutation-erasing treatments in the long run.

Despite the fact that humans might not be rapidly evolving on a species level, the forces of competition still operate inside us among our cells, as Wilhelm Roux described in the 1800s. Roux considered the battle between components of organisms part of an ongoing struggle since life formed on Earth. "The cultivating struggle between the living parts, evoked by the differences between them, will . . . have begun with the first origin of life and never ceased since then," he explained.

We have awakened to the evolution happening within us, and we have to be very careful about how future medicines might interfere with this process. And we need to be realistic about how much we can actually curb it. Roux's wise words are helpful here. "Nothing can be kept absolutely constant because everything constantly changes," he wrote. Roux did not know about DNA when he published those words, but in the century and a half since then, we have discovered that mutation can have positive, negative, and even neutral consequences. As the geneticist and historian Elof Axel Carlson has put it, "Mutation, of course, involves change, but our understanding of that change is influenced by the time we live in."

The more we learn about the genetic code that runs our cells, the more we appreciate that it is a dynamic set of instructions. There are changes big and small that happen to the DNA sequences within our tissues as we grow and age. There is a bit of chaos in our genome. It's mind-bending to think about the endless permutations of genetic tweaks that occur naturally by chance. But there might also be a way to embrace this messy reality. Rather than seek to put an end to all mutations full stop, we should welcome the helpful ones and accept the harmless ones as a part of who we are. It's time that we acknowledge the vast sequence diversity that emerges within each of us. As we move through the world, our bodies are brimming with genetic possibilities that have the power to shape our future.

Acknowledgments

For several years, I carried around the idea for this book with me without knowing how to bring it forward. A few key conversations gave me the confidence to keep pushing ahead. One of them happened during a short walk with the author Charles Seife on a spring day in New York in 2017. His enthusiasm about the concept of this book gave me the guts to stick with it even when well-meaning people tried pulling me in other directions. Charles and I spoke so briefly, but our exchange had a lasting influence on my desire to see this book idea through.

Not too long after that conversation, I got to meet Robert Butler in Florida and talk with him about his cancer diagnosis and his participation as one of the first people in an adaptive therapy trial. Together we attended a meeting of scientists brainstorming how to tackle his type of disease by leveraging the principle of evolution. But Robert did not sit by the sidelines. He was actively involved in trying to help doctors understand how influencing mutation in the body could help patients like him. Robert convinced me that other patients would be curious about this new field of medicine.

I'm forever grateful to people such as Robert who let me into their lives and shared their experiences with me. Sometimes we communicated in moments when things were most difficult and uncertain, and their grace and courage left me awestruck on many occasions. Thank you, truly, to all the patients and their families whose names appear in the pages of this book.

There are many individuals who gave me their time in interviews and emails but whose names I have not mentioned until now. This is my opportunity to thank them for helping shape this book from behind the scenes. I am grateful to Jamie Butler, Carolyn Butler, Christopher Griffin, Noah Davidsohn, Mami Taniuchi, Dawn Lemanne, Kelly Bolton, Neal Young, Junko Oshima, Hans Ochs, Nicholas E. Baker, Patrick O'Farrell, and Brian Gear. While he is quoted briefly in this book, Mike McConnell provided vital input over several years for the sections related to the brain and immunity. Thank you also to Claudia Steinicke, who steered me toward crucial documents pertaining to the life of Wilhelm Roux, and to Karen Schindler, who reviewed several key passages related to reproductive cells. Cassius Adair gave me essential guidance on inclusive language pertaining to parentage, gender, and inheritance. Gerald Simon and Fabian Garcia provided valuable expertise on Nietzsche. It's also important for me to acknowledge the help of Andrea Vogt, Carolin Wett, and Jan Doering, who all guided me on German translations.

Anyone who attempts to write a book while parenting a toddler will tell you it takes a whole lot of assists to make that possible. I owe a tremendous amount to the wonderful women who helped me during these years: Chloé Fostier, Arman Doosti, Charlène Gueguen, Ava Kasianchuk, Alice Dautigny, and Sarah Beydoun. Sarah also had a huge role in shaping the title of the book and giving helpful feedback on early passages as well as the final read-through. Thanks also goes to my cousin Natalie Pavlovic, who lent a hand with childcare and came with useful reference materials on genetics and immunity in tow.

I'm grateful to the Alfred P. Sloan Foundation for its generous support, and to *MIT Technology Review* and *Wired* magazine for publishing some of my writing about genetic mutation that seeded this book. In particular, I'm indebted to Vera Titunik for her edits on my *Wired* feature story "The Darwin Treatment," which detailed the genesis of adaptive therapy for cancer.

Michael Pollan, who has given me writing advice at pivotal moments throughout the past decade, helped me figure out how to refine my book proposal and navigate the terrain of book writing for the first time. Another person who pointed me in the right direction many times in this process is Ed Yong.

As I embarked on this project, I was delighted to find a new friend in Julia Belluz. Julia was a few months ahead of me on her book-writing timeline and gave me indispensable tips about what to anticipate down the pike. I've also benefited from book-related guidance in these past few years from brilliant authors such as Bethany Brookshire, Maryn McKenna, Adam Rogers, Christie Aschwanden, Rebecca Boyle, Kendra Pierre-Louis, Alan Henry, Maggie Koerth, and Brooke Borel. Additionally, I'm grateful to Dan Engber and Emily Laber-Warren for all they've taught me about health journalism during the past half decade when I was formulating this book. My ability to find important research papers and translate complicated science was buoyed by my former TEDMED advisory board members Kafui Dzirasa, Céline Gounder, Giles Newton, and Udaya Patnaik, as well as Shirley Bergin. Amy Maxmen, a gifted journalist whom I am lucky to call a dear friend, was always there to brainstorm with me at many reporting crossroads.

As I neared the end of writing this book, two scientists graciously took time out of their busy schedules to closely review the near-final draft and provide comments on the text. Lynn Caporale, biochemist and author of *Darwin in the Genome*, offered enormously insightful feedback, including about the history of evolutionary thought and genetic sequencing, and Jeffrey Townsend, a professor at the Yale School of Public Health and cochair of the American Association for Cancer Research's Cancer Evolution Working Group, gave detailed notes to sharpen the accuracy of the wording around many crucial points, including passages related to sequencing technology and the intricacies of mutations in tumors.

I had the tremendous good fortune to work with Brad Scriber, who not only fact-checked this book but also gave shrewd recommendations on how to make the science in its pages more accessible to readers. Brad never shied from digging into the subtleties of a study and turned up many important truths to elevate the accuracy of my sentences. His attention to detail surpasses that of anyone I have ever met.

My editor at Riverhead Books, Courtney Young, is the kind of person every author dreams of working with: someone who thoughtfully clarifies and amplifies the message you are trying to convey while preserving its essence. At every step of the way she kindly offered course corrections and asked all the right questions to keep things on a forward path. Courtney took my drafts, which were overflowing with scientific detail, and knew exactly which bits to trim. She gave so many hours to this book and saw it through various structural iterations. I'm thankful for all her patience, and for her precision.

At the end of the day, this book would not exist without my agent, Will Francis. Will quickly grasped my goal of writing an intellectual history that spanned multiple centuries and scientific disciplines—and, unsurprisingly to anyone who knows him, he immediately articulated this vision even better than I could myself. Will selflessly helped shape this book at each stage. He lent his knowledge about how evolution and genetics have been conveyed in the past, and found ways to emphasize the inflection point we are at now that DNA sequencing has revealed that each of us acquires countless mutations. Will helped me lay the foundation for the book and build it from there, and was present in its formation even though he was many time zones away.

Finally, I need to thank my parents, who have nourished my love of books and science for decades. My mother, Ladan, taught me to be a writer, and my father, Farhad, taught me to be a reader. When I was young, they filled the bookcases in our house with the latest bestsellers about biology and physics and would gather us around the television to watch the *Nova* series about new developments in research. In the past few years, they have gone the extra mile and often did daycare drop-offs and pickups so that I could stay parked at the computer and finish chapters. They endured my obsession with the topic of mutation and took every opportunity to tell me they were counting down the days until they could read the book I was working on. I hope they can finally enjoy it now.

Notes

INTRODUCTION

ix **At the end of this:** C. G. Petersen et al., "Embryo Selection by the First Cleavage Parameter Between 25 and 27 Hours After ICSI," *Journal of Assisted Reproduction and Genetics* 18, no. 4 (2001): 209–12, doi.org/10.1023/a:100946 0013579.

ix **a cluster of sixteen cells:** Jie Wang et al., "The Influence of Day 3 Embryo Cell Number on the Clinical Pregnancy and Live Birth Rates of Day 5 Single Blastocyst Transfer from Frozen Embryo Transfer Cycles," *BMC Pregnancy and Childbirth* 22, no. 1 (2022), doi.org/10.1186/s12884-022-05337-z.

ix **You graduated from this:** Tijana Vlajkovic et al., "Day 5 versus Day 3 Embryo Biopsy for Preimplantation Genetic Testing for Monogenic/Single Gene Defects," *Cochrane Database of Systematic Reviews* 2022, no. 11 (2022), doi.org /10.1002/14651858.cd013233.pub2.

ix **Ultimately, your body developed:** Yella Hewings-Martin, "How Many Cells Are in the Human Body?," *Medical News Today*, July 12, 2017, medicalnews today.com/articles/318342.

x **this theorized organism:** Marcelo Gleiser, "The Microbial Eve: Our Oldest Ancestors Were Single-Celled Organisms," *National Public Radio*, January 31, 2018, npr.org/sections/13.7/2018/01/31/581874421/be-humbled-our-oldest-an cestors-were-single-celled-organisms.

x **more than ten times:** Karl J. Niklas and Stuart A. Newman, "The Many Roads to and from Multicellularity," *Journal of Experimental Botany* 71, no. 11 (2020): 3247–53, doi.org/10.1093/jxb/erz547.

x **German naturalist Ernst Haeckel:** Jordana Cepelewicz, "Scientists Debate the Origin of Cell Types in the First Animals," *Quanta Magazine*, July 17, 2019,

quantamagazine.org/scientists-debate-the-origin-of-cell-types-in-the-first
-animals-20190717.

xi   **One tragic case:** Kat Arney, *Rebel Cell: Cancer, Evolution, and the New Science
of Life's Oldest Betrayal* (BenBella Books, 2020), 262.

xi   **Then there's a case:** Hermine-Valeria Gärtner et al., "Genetic Analysis of a
Sarcoma Accidentally Transplanted from a Patient to a Surgeon," *New England
Journal of Medicine* 335, no. 20 (1996): 1494–97, doi.org/10.1056/nejm1996111
43352004.

xii   **The concept of mutation emerged:** Masatoshi Nei and Masafumi Nozawa,
"Roles of Mutation and Selection in Speciation: From Hugo de Vries to the
Modern Genomic Era," *Genome Biology and Evolution* 3 (January 2011):
812–29, doi.org/10.1093/gbe/evr028.

xii   **some seeds from the patch:** Siddhartha Mukherjee, *The Gene: An Intimate
History* (Scribner, 2016), 60–61.

xii   **He considered these different species:** Hugo de Vries, *Species and Varieties:
Their Origin by Mutation*, 2nd ed. (Open Court, 1906), 562.

xiii   **One of the first inklings:** Andrew J. Holland and Don W. Cleveland, "Boveri
Revisited: Chromosomal Instability, Aneuploidy and Tumorigenesis," *Nature
Reviews: Molecular Cell Biology* 10, no. 7 (2009): 478–87, doi.org/10.1038
/nrm2718.

xiii   **with an unbalanced number:** Manfred Dietel, "Boveri at 100: The Life and
Times of Theodor Boveri," *Journal of Pathology* 234, no. 2 (2014): 135–37,
doi.org/10.1002/path.4410.

xiii   **In 1914, the same year:** Robert A. Weinberg, "In Retrospect: The Chromosome Trail," *Nature* 453, no. 7196 (2008): 725, doi.org/10.1038/453725a.

xiii   **In 1976, he proposed:** P. C. Nowell, "The Clonal Evolution of Tumor Cell
Populations," *Science* 194, no. 4260 (1976): 23–28, doi.org/10.1126/science
.959840.

xiii   **"We are here to celebrate":** "Remarks by the President . . . on the Completion of the Entire Human Genome Project," White House Office of the Press
Secretary, June 26, 2000, clintonwhitehouse4.archives.gov/WH/EOP/OSTP
/html/00628_2.html.

xiv   **the two teams working:** Gavin Yamey, "Scientists Unveil First Draft of
Human Genome," *BMJ: British Medical Journal* 321, no. 7252 (2000): 7, doi
.org/10.1136/bmj.321.7252.7.

xiv   **an impressive 92 percent:** "First Complete Sequence of a Human Genome,"
National Institutes of Health, accessed July 19, 2025, nih.gov/news-events/nih
-research-matters/first-complete-sequence-human-genome.

xiv   **an end-to-end, gapless sequence:** Prabarna Ganguly and Rachael Zisk,

"Researchers Generate the First Complete, Gapless Sequence of a Human Genome," National Human Genome Research Institute, March 31, 2022, genome.gov/news/news-release/researchers-generate-the-first-complete-gapless -sequence-of-a-human-genome.

xiv **This included the Y chromosome:** Arang Rhie et al., "The Complete Sequence of a Human Y Chromosome," *Nature* 621, no. 7978 (2023): 344–54, doi.org /10.1038/s41586-023-06457-y.

xiv **at least twenty thousand genes:** "What Is a Gene?," MedlinePlus, updated May 21, 2024, medlineplus.gov/genetics/understanding/basics/gene.

xiv **are around 99.5 percent similar:** "Fact Sheet: Human Genomic Variation," National Human Genome Research Institute, accessed January 14, 2025, ge nome.gov/about-genomics/educational-resources/fact-sheets/human-genomic -variation; Adam Auton and Gonçalo R. Abecasis, "A Global Reference for Human Genetic Variation," *Nature* 526, no. 7571 (2015): 68–74, doi.org /10.1038/nature15393.

xiv **about 98.4 percent identical:** Kelly A. Frazer et al., "Genomic DNA Insertions and Deletions Occur Frequently Between Humans and Nonhuman Primates," *Genome Research* 13, no. 3 (2003): 341–46, doi.org/10.1101/gr.554603.

xv **was the first to be identified:** Stylianos E. Antonarakis, "History of the Methodology of Disease Gene Identification," *American Journal of Medical Genetics, Part A* 185, no. 11 (2021): 3266–75, doi.org/10.1002/ajmg.a.62400.

xv **identified by scientists skyrocketed:** Melina Claussnitzer et al., "A Brief History of Human Disease Genetics," *Nature* 577, no. 7789 (2020): 179–89, doi .org/10.1038/s41586-019-1879-7.

xv **seven thousand inherited ailments:** Stephen F. Kingsmore, Russell Nofsinger, and Kasia Ellsworth, "Rapid Genomic Sequencing for Genetic Disease Diagnosis and Therapy in Intensive Care Units: A Review," *NPJ Genomic Medicine* 9, no. 1 (2024): 17, doi.org/10.1038/s41525-024-00404-0.

xv **we have amassed:** Jan Vijg and Xiao Dong, "Pathogenic Mechanisms of Somatic Mutation and Genome Mosaicism in Aging," *Cell* 182, no. 1 (2020): 12–23, doi.org/10.1016/j.cell.2020.06.024.

xvi **In 2009, scientists announced:** Dominic Grün and Alexander van Oudenaarden, "Design and Analysis of Single-Cell Sequencing Experiments," *Cell* 163, no. 4 (2015): 799–810, doi.org/10.1016/j.cell.2015.10.039.

xvi **Further methods devised:** Bora Lim, Yiyun Lin, and Nicholas Navin, "Advancing Cancer Research and Medicine with Single-Cell Genomics," *Cancer Cell* 37, no. 4 (2020): 456–70, doi.org/10.1016/j.ccell.2020.03.008.

xvi **Breakthrough of the Year:** Elizabeth Pennisi, "2018 Breakthrough of the Year," *Science*, December 20, 2018, science.org/content/article/breakthrough -2018/finalists.

xviii **Smoking just fifteen:** Erin D. Pleasance et al., "A Small Cell Lung Cancer Genome with Complex Tobacco Exposure Signatures," *Nature* 463, no. 7278 (2010): 184–90, doi.org/10.1038/nature08629.

xviii **Every cell in sun-exposed skin:** "Using Healthy Skin to Identify Cancer's Origins," Wellcome Sanger Institute, May 21, 2025, sanger.ac.uk/news_item /2015-05-21-using-healthy-skin-to-identify-cancer-s-origins.

xviii **approximately three billion:** A. J. Marian, "Sequencing Your Genome: What Does It Mean?," *Methodist DeBakey Cardiovascular Journal* 10, no. 1 (2014): 3–6, doi.org/10.14797/mdcj-10-1-3.

xviii **These enzymes can introduce:** Emily K. Law et al., "APOBEC3A Catalyzes Mutation and Drives Carcinogenesis In Vivo," *Journal of Experimental Medicine* 217, no. 12 (2020): e20200261, doi.org/10.1084/jem.20200261; Michael Morse et al., "HIV Restriction Factor APOBEC3G Binds in Multiple Steps and Conformations to Search and Deaminate Single-Stranded DNA," *eLife* 8 (December 2019): e52649, doi.org/10.7554/eLife.52649.

xviii **colon and liver tumors:** Law et al., "APOBEC3A Catalyzes Mutation."

xviii **more than twenty human cancers:** Arney, *Rebel Cell*, 80.

xix **Jan Vijg and Xiao Dong have noted:** Vijg and Dong, "Pathogenic Mechanisms of Somatic Mutation."

xix **that replenishes tissues:** Arney, *Rebel Cell*, 57.

xix **more than three thousand mutations:** Sarah Zhang, "Your Body Acquires Trillions of New Mutations Every Day," *Atlantic*, May 7, 2018, theatlantic.com /science/archive/2018/05/your-body-acquires-trillions-of-new-mutations -every-day/559472.

xx **a few scientists even speculate:** Kimberlee D'Ardenne, "'Big Bang' Model of Colon Cancer Identifies Role Time Plays in Tumor-Growth Dynamics," Stanford Medicine, February 10, 2015, med.stanford.edu/news/all-news/2015-02 /big-bang-model-of-colon-cancer-identifies-role-time-plays.html.

xxiv **stunning number of mutations:** Carl Zimmer, "Every Cell in Your Body Has the Same DNA. Except It Doesn't," *New York Times*, May 21, 2018, nytimes .com/2018/05/21/science/mosaicism-dna-genome-cancer.html.

xxiv **By the team's calculations:** Michael A. Lodato et al., "Somatic Mutation in Single Human Neurons Tracks Developmental and Transcriptional History," *Science* 350, no. 6256 (2015): 94–98, doi.org/10.1126/science.aab1785.

xxiv **"a beautiful and unique snowflake":** Ed Yong, "The Surprising Genealogy of Your Brain," *Atlantic*, October 1, 2015, theatlantic.com/science/archive/2015 /10/the-genealogy-of-your-brain/408232.

xxiv **"their own worst enemy":** Pamela J. Hines, "This Week in Science," *Science* 350, no. 6256 (2015): 52–54, doi.org/10.1126/science.2015.350.6256.twis.

xxiv **"slowly but inexorably with age"**: Michael A. Lodato et al., "Aging and Neurodegeneration Are Associated with Increased Mutations in Single Human Neurons," *Science* 359, no. 6375 (2018): 555–59, doi.org/10.1126/science.aao4426.

xxiv **twenty new mutations per year**: Michael B. Miller, Hannah C. Reed, and Christopher A. Walsh, "Brain Somatic Mutation in Aging and Alzheimer's Disease," *Annual Review of Genomics and Human Genetics* 22 (August 2021): 239–56, doi.org/10.1146/annurev-genom-121520-081242.

xxiv **British researchers calculated**: Michael J. Keogh et al., "High Prevalence of Focal and Multi-Focal Somatic Genetic Variants in the Human Brain," *Nature Communications* 9, no. 1 (2018): 4257, doi.org/10.1038/s41467-018-06331-w.

xxiv **to one million brain cells**: Jessica Shugart, "Islands of Mutated Neurons Dot the Brain. Are They Bad for Us?," *Alzforum*, October 23, 2018, alzforum.org/news/research-news/islands-mutated-neurons-dot-brain-are-they-bad-us.

xxv **possess genetic variations**: Long Wang et al., "The Architecture of Intra-Organism Mutation Rate Variation in Plants," *PLOS Biology* 17, no. 4 (2019): e3000191, doi.org/10.1371/journal.pbio.3000191.

CHAPTER 1: TURNING CANCER AGAINST ITSELF

1 **"Every one admits"**: Charles Darwin, *The Variation of Animals and Plants Under Domestication* (John Murray, 1875), 366, darwin-online.org.uk/converted/published/1875_Variation_F880/1875_Variation_F880.2.html.

1 **Robert Butler moved to Tampa**: Roxanne Khamsi, "The Darwin Treatment," *Wired*, April 2019, 71.

3 **"The misery I endured"**: Larry A. Nielsen, "Second Voyage of the Beagle Began (1831)," *Today in Conservation*, accessed March 8, 2023, todayinconservation.com/2020/01/december-27-second-voyage-of-the-beagle-began-1831.

4 **15 percent died from the treatment itself**: Daniel F. Hayes, "False Hope: Bone Marrow Transplantation for Breast Cancer," *New England Journal of Medicine* 357, no. 10 (2007): 1059–60, doi.org/10.1056/NEJMbkrev58584.

6 ***Cancer Research* in 1991**: Robert A. Gatenby, "Population Ecology Issues in Tumor Growth," *Cancer Research* 51, no. 10 (1991): 2542–47.

7 **The tumors he mentions**: Darwin, *Variation of Animals and Plants*, 366.

7 **Lawson Tait, a gynecologist**: J. A. Shepherd, "Lawson Tait: Disciple of Charles Darwin," *British Medical Journal (Clinical Research Ed.)* 284, no. 6326 (1982): 1386, doi.org/10.1136/bmj.284.6326.1386.

7 **"Mr. Lawson Tait refers to a tumour"**: Darwin, *Variation of Animals and Plants*, 366.

8   **tugging the material in uneven ways:** Kat Arney, *Rebel Cell: Cancer, Evolution, and the New Science of Life's Oldest Betrayal* (BenBella Books, 2020), 64.

9   **"it appears logical to regard":** Ernest E. Tyzzer, "Tumor Immunity," *Journal of Cancer Research* 1, no. 2 (1916): 125–56.

9   **"it is conceivable":** Thomas Hunt Morgan and Calvin Bridges, *Contributions to the Genetics of Drosophila melanogaster* (Carnegie Institution of Washington, 1919), 109.

9   **The idea that tumors:** Arney, *Rebel Cell*, 66.

9   **Some even thought:** Peter C. Nowell, "Discovery of the Philadelphia Chromosome: A Personal Perspective," *Journal of Clinical Investigation* 117, no. 8 (2007): 2033–35, doi.org/10.1172/JCI31771.

10   **he saw a crucial commonality:** Peter C. Nowell, "Tumor Progression: A Brief Historical Perspective," *Seminars in Cancer Biology* 12, no. 4 (2002): 261–66, doi.org/10.1016/S1044-579X(02)00012-3.

10   **in a seminal 1976 paper:** Peter C. Nowell, "The Clonal Evolution of Tumor Cell Populations," *Science* 194, no. 4260 (1976): 23–28, doi.org/10.1126/science .959840.

11   **Around 9 percent:** B. Vogelstein et al., "Genetic Alterations During Colorectal-Tumor Development," *New England Journal of Medicine* 319, no. 9 (1988): 525–32, doi.org/10.1056/NEJM198809013190901.

11   **This ultimately gave rise:** Eric R. Fearon and Bert Vogelstein, "A Genetic Model for Colorectal Tumorigenesis," *Cell* 61, no. 5 (1990): 759–67, doi .org/10.1016/0092-8674(90)90186-i.

12   **"literally riddled with holes":** Asa Fitch, *Noxious, Beneficial and Other Insects of the State of New York* (C. Van Benthuysen, 1856), 171.

16   **They published a report in 2017:** Jingsong Zhang et al., "Integrating Evolutionary Dynamics into Treatment of Metastatic Castrate-Resistant Prostate Cancer," *Nature Communications* 8, no. 1 (2017): 1816, doi.org/10.1038/s41467 -017-01968-5.

16   **the results looked good overall:** Jingsong Zhang et al., "Evolution-Based Mathematical Models Significantly Prolong Response to Abiraterone in Metastatic Castrate-Resistant Prostate Cancer and Identify Strategies to Further Improve Outcomes," *eLife* 11 (June 2022): e76284, doi.org/10.7554/eLife .76284.

18   **A significant experiment published in 2012:** Marco Gerlinger et al., "Intratumor Heterogeneity and Branched Evolution Revealed by Multiregion Sequencing," *New England Journal of Medicine* 366, no. 10 (2012): 883–92, doi .org/10.1056/nejmoa1113205.

18   **double most of the genome:** "Cancer's 'Genome Doubling' Mystery Solved,"

Francis Crick Institute, March 6, 2020, crick.ac.uk/news/2020-03-06_cancers
-genome-doubling-mystery-solved.

19 **but often spur resistance mutations:** Arney, *Rebel Cell*, 176.

19 **At the end of the experiment:** Jill A. Gallaher et al., "Spatial Heterogeneity
and Evolutionary Dynamics Modulate Time to Recurrence in Continuous
and Adaptive Cancer Therapies," *Cancer Research* 78, no. 8 (2018): 2127–39,
doi.org/10.1158/0008-5472.CAN-17-2649.

20 **study of colon cancer samples:** Andrea Sottoriva et al., "A Big Bang Model of
Human Colorectal Tumor Growth," *Nature Genetics* 47, no. 3 (2015): 209–16,
doi.org/10.1038/ng.3214.

20 **imploring the president:** Citizens Committee for the Conquest of Cancer,
"Mr. Nixon: You Can Cure Cancer," *Washington Post*, December 9, 1969, C9.

22 **coauthored a science paper:** Renee Brady-Nicholls et al., "Predicting Patient-
Specific Response to Adaptive Therapy in Metastatic Castration-Resistant
Prostate Cancer Using Prostate-Specific Antigen Dynamics," *Neoplasia* 23,
no. 9 (2021): 851–58, doi.org/10.1016/j.neo.2021.06.013.

23 **bigger adaptive therapy trial:** "Personalised Adaptive Treatment for Prostate
Cancer: New Clinical Trial Enrols First Participants," *Oncology News*, March
7, 2023, oncologynews.com.au/latest-news/personalised-adaptive-treatment
-for-prostate-cancer-new-clinical-trial-enrolls-first-participants.

24 **from 3 percent to 90 percent:** Arney, *Rebel Cell*, 242.

25 **increases tumor cells' "evolvability":** Kenneth J. Pienta et al., "Poly-
aneuploid Cancer Cells Promote Evolvability, Generating Lethal Cancer," *Evolutionary Applications* 13, no. 7 (2020): 1626–34, doi.org/10.1111/eva.12929.

25 **sometimes lie dormant:** Arney, *Rebel Cell*, 175–76.

25 **patients with medulloblastoma:** A. Sorana Morrissy et al., "Divergent Clonal
Selection Dominates Medulloblastoma at Recurrence," *Nature* 529, no. 7586
(2016): 351–57, doi.org/10.1038/nature16478.

26 **five-square-centimeter patch:** Arney, *Rebel Cell*, 108–9.

26 **there were more mutations:** Iñigo Martincorena et al., "Somatic Mutant
Clones Colonize the Human Esophagus with Age," *Science* 362, no. 6417
(2018): 911–17, doi.org/10.1126/science.aau3879.

26 **large parts of the esophagus:** "Protective Mutation Impairs Oesophagus Tumour Growth," Wellcome Sanger Institute, January 19, 2023, sanger.ac.uk
/news_item/protective-mutation-impairs-oesophagus-tumour-growth.

26 **in the rodents:** Emilie Abby et al., "*Notch1* Mutations Drive Clonal Expansion
in Normal Esophageal Epithelium but Impair Tumor Growth," *Nature Genetics* 55, no. 2 (2023): 232–45, doi.org/10.1038/s41588-022-01280-z.

26 **"From the point of view":** Kamila Naxerova, "Mutation Fingerprints Encode

Cellular Histories," *Nature* 597, no. 7876 (2021): 334–36, doi.org/10.1038 /d41586-021-02269-0.

27 **one in ten people:** Diego F. Coutinho, Ilana R. Zalcberg, and Bárbara C. R. Monte-Mór, "Myeloid Malignancies-Related Somatic Mutations in Aging Individuals," *Molecular Genetics & Genomic Medicine* 7, no. 6 (2019): e683, doi .org/10.1002/mgg3.683.

27 **can be kept at bay:** Arney, *Rebel Cell*, 135.

27 **experiments by embryologists:** Mina J. Bissell and William C. Hines, "Why Don't We Get More Cancer? A Proposed Role of the Microenvironment in Restraining Cancer Progression," *Nature Medicine* 17, no. 3 (2011): 320–29, doi.org/10.1038/nm.2328.

27 **"He took the paper":** Gina Kolata, "Old Ideas Spur New Approaches in Cancer Fight," *New York Times*, December 28, 2009, nytimes.com/2009/12/29 /health/research/29cancer.html.

28 **coaxed to become tame:** V. M. Weaver et al., "Reversion of the Malignant Phenotype of Human Breast Cells in Three-Dimensional Culture and In Vivo by Integrin Blocking Antibodies," *Journal of Cell Biology* 137, no. 1 (1997): 231–45, doi.org/10.1083/jcb.137.1.231; Arney, *Rebel Cell*, 135.

CHAPTER 2: THE STRUGGLE IS REAL

29 **"Uniformity is pure delirium":** Friedrich Nietzsche, *Sämtliche Werke: Kritische Studienausgabe, Bd. 9* (Deutscher Taschenbuch Verlag; De Gruyter, 1980), 490.

29 **born to a family:** Alex Kiermayer, "The Evolution of German Cut Fencing in the 19th Century Viewed Through the Works of Friedrich August Wilhelm Ludwig Roux," *Acta Periodica Duellatorum* 6, no. 2 (2020): 77–102, doi.org /10.36950/apd-2018-008.

29 **his father expected:** Wilhelm Roux, *The Struggle of Parts*, trans. David Haig and Richard Bondi (Harvard University Press, 2024), 4.

29 **Conflict surrounded Wilhelm Roux:** Roux, *The Struggle of Parts*, 3.

30 **Roux described himself:** Ulrike Feicht, "Wilhelm Roux (1850–1924)—seine hallesche Zeit" (DMD diss., University of Halle-Wittenberg, 2008), 8, opendata .uni-halle.de/bitstream/1981185920/10576/1/prom.pdf.

30 **he had a "*freudearm*":** Frederick B. Churchill, "Roux, Wilhelm," in *Dictionary of Scientific Biography*, ed. Charles Coulston Gillispie (Scribner, 1981), 570–74.

30 **Herbert Spencer was inspired:** Dan Falk, "The Complicated Legacy of Herbert Spencer, the Man Who Coined 'Survival of the Fittest,'" *Smithsonian*,

April 29, 2020, smithsonianmag.com/science-nature/herbert-spencer-survival -of-the-fittest-180974756.

31 **the urging of fellow naturalist:** Alfred Russel Wallace, letter to Charles Darwin, July 2, 1866, darwinproject.ac.uk/letter/?docId=letters/DCP-LETT-5140.xml.

31 **He studied for several years:** Feicht, "Wilhelm Roux (1850–1924)," 4.

31 **Virchow's work echoed:** David Lagunoff, "A Polish, Jewish Scientist in 19th-Century Prussia," *Science* 298, no. 5602 (2002): 2331, doi.org/10.1126/science .1080726.

31 **Roux was swayed by his mentor:** Thomas Heams, "Selection Within Organisms in the Nineteenth Century: Wilhelm Roux's Complex Legacy," *Progress in Biophysics and Molecular Biology* 110, no. 1 (2012): 24–33, doi.org/10.1016 /j.pbiomolbio.2012.04.004.

31 **Roux was convinced:** Alessandra Passariello, "Wilhelm Roux (1850–1924) and Blood Vessel Branching," *Journal of Theoretical and Applied Vascular Research* 1, no. 1 (2016), 25–40, doi.org/10.24019/jtavr.4.

32 **inspired by the work:** Bartlomiej Swiatczak, "Evolution Within the Body: The Rise and Fall of Somatic Darwinism in the Late Nineteenth Century," *History and Philosophy of the Life Sciences* 45, no. 1 (2023), doi.org/10.1007 /s40656-023-00566-7.

32 **tongue pushing against the teeth:** Swiatczak, "Evolution Within the Body."

33 **first few pages of *Der Kampf*:** Heams, "Selection Within Organisms"; Roux, *The Struggle of Parts*, 88.

33 **occurring on four levels:** Swiatczak, "Evolution Within the Body."

33 **When Roux referred to "molecules":** Heams, "Selection Within Organisms."

34 **"Because, as we have seen":** Roux, *The Struggle of Parts*, 107.

34 **As Roux wrote in *Der Kampf*:** Heams, "Selection Within Organisms"; Roux, *The Struggle of Parts*, 38.

35 **Nietzsche used what he learned:** Swiatczak, "Evolution Within the Body."

35 **Thomas Heams, a biologist:** Heams, "Selection Within Organisms."

35 **who was quite taken:** Charles Darwin, letter to George John Romanes, *The Life and Letters of George John Romanes* (Longmans, Green, 1908), 115, darwin -online.org.uk/content/frameset?itemID=F2111&viewtype=text&pageseq=1.

36 **"Perhaps the most striking feature":** Roux, *The Struggle of Parts*, 2.

36 **as explicitly and thoroughly:** Swiatczak, "Evolution Within the Body."

36 **"That Roux had a substantial impact":** Churchill, "Roux, Wilhelm."

37 **In one famous experiment:** Edward M. De Robertis, "Spemann's Organizer and Self-Regulation in Amphibian Embryos," *Nature Reviews: Molecular Cell*

Biology 7, no. 4 (2006): 296–302, doi.org/10.1038/nrm1855; Viktor Hamburger, "Wilhelm Roux: Visionary with a Blind Spot," *Journal of the History of Biology* 30, no. 2 (1997): 229–38.

37 **Swiatczak points to a movement:** Swiatczak, "Evolution Within the Body."

38 **evolutionary dead ends:** Swiatczak, "Evolution Within the Body."

38 **he would walk every day at seven a.m.:** Feicht, "Wilhelm Roux (1850–1924)," 8.

38 **when two PhD students in Spain:** G. Morata and P. Ripoll, "Minutes: Mutants of Drosophila Autonomously Affecting Cell Division Rate," *Developmental Biology* 42, no. 2 (1975): 211–21, doi.org/10.1016/0012-1606(75)90330-9; Kendall Powell, "Survival of the Fittest Cells," *Nature* 574, no. 7778 (2019): 310–12, doi.org/10.1038/d41586-019-03060-y.

39 **"Even though there was":** Ginés Morata, "Cell Competition: A Historical Perspective," *Developmental Biology* 476 (August 2021): 33–40, doi.org/10.1016/j.ydbio.2021.02.012.

39 **This is what the scientists found:** Stephano Mello and Dirk Bohmann, "Cell Competition: Counting the Minutes," *eLife* 9 (January 2020), doi.org/10.7554/elife.53348.

40 **Cells with mutations can be eliminated:** Marisa M. Merino, Romain Levayer, and Eduardo Moreno, "Survival of the Fittest: Essential Roles of Cell Competition in Development, Aging, and Cancer," *Trends in Cell Biology* 26, no. 10 (2016): 776–88, doi.org/10.1016/j.tcb.2016.05.009.

40 **have discovered more than a dozen:** Kara L. McKinley, David Castillo-Azofeifa, and Ophir D. Klein, "Tools and Concepts for Interrogating and Defining Cellular Identity," *Cell Stem Cell* 26, no. 5 (2020): 632–56, doi.org/10.1016/j.stem.2020.03.015.

40 **super-fit mutant heart cells:** Sarah Bowling, Katerina Lawlor, and Tristan A. Rodriguez, "Cell Competition: The Winners and Losers of Fitness Selection," *Development* 146, no. 13 (2019), doi.org/10.1242/dev.167486.

40 **those that were "unfit":** Margarida Sancho et al., "Competitive Interactions Eliminate Unfit Embryonic Stem Cells at the Onset of Differentiation," *Developmental Cell* 26, no. 1 (2013): 19–30, doi.org/10.1016/j.devcel.2013.06.012.

41 **fertility specialists from New York reported:** N. Gleicher et al., "Further Evidence Against Use of PGS in Poor Prognosis Patients: Report of Normal Births after Transfer of Embryos Reported as Aneuploid," *Fertility and Sterility* 104, no. 3 (2015): e59, doi.org/10.1016/j.fertnstert.2015.07.180.

41 **doctors in Italy detailed:** Ermanno Greco, Maria Giulia Minasi, and Francesco Fiorentino, "Healthy Babies After Intrauterine Transfer of Mosaic Aneuploid Blastocysts," *New England Journal of Medicine* 373, no. 21 (2015): 2089–90, doi.org/10.1056/nejmc1500421.

41 **preimplantation tests might pick up:** Azeen Ghorayshi, "Study Raises Questions About Popular Genetic Test for 'Abnormal' Embryos," Health, *New York Times*, April 20, 2022, nytimes.com/2022/04/20/health/pgta-ivf-pregnancy-test.html.

42 **no two cells within the same organ:** Heams, "Selection Within Organisms."

43 **a third to two-thirds of the total:** Bruce Alberts et al., "An Overview of Gene Control," in *Molecular Biology of the Cell*, 4th ed. (Garland Science, 2002), ncbi.nlm.nih.gov/books/NBK26885.

43 **The seeds of this project:** Anna Nowogrodzki, "The Cell Seeker," *Nature* 547, no. 7661 (2017): 24–26, doi.org/10.1038/547024a.

44 **"the time is ripe to complete":** Aviv Regev et al., "Science Forum: The Human Cell Atlas," *eLife* 6 (2017): e27041, doi.org/10.7554/eLife.27041.

45 **The average age of the cells:** Nicholas Wade, "Your Body Is Younger Than You Think," *New York Times*, August 2, 2005, nytimes.com/2005/08/02/science/your-body-is-younger-than-you-think.html.

45 **There are cells that hang around:** Roxanne Khamsi, "Carbon Dating Works for Cells," *Nature*, July 14, 2005, doi.org/10.1038/news050711-12.

45 **a thousand skin cells:** D. Roberts and R. Marks, "The Determination of Regional and Age Variations in the Rate of Desquamation: A Comparison of Four Techniques," *Journal of Investigative Dermatology* 74, no. 1 (1980): 13–16, doi.org/10.1111/1523-1747.ep12514568.

45 **the body replaces 330 billion:** Mark Fischetti and Jen Christiansen, "Our Bodies Replace Billions of Cells Every Day," *Scientific American*, April 1, 2021, scientificamerican.com/article/our-bodies-replace-billions-of-cells-every-day.

47 **"Every human is undoubtedly mosaic":** Ian M. Campbell et al., "Somatic Mosaicism: Implications for Disease and Transmission Genetics," *Trends in Genetics* 31, no. 7 (2015): 382–92, doi.org/10.1016/j.tig.2015.03.013.

48 **the fittest as endoevolution:** Roxanne Khamsi, "The Darwin Treatment," *Wired*, April 2019, 71.

48 **once promised to donate his brain:** Feicht, "Wilhelm Roux (1850–1924)," 10.

48 **for his body to be injected:** Feicht, "Wilhelm Roux (1850–1924)," 10.

49 **"stability arising through struggle":** Stephen Jay Gould, *The Structure of Evolutionary Theory* (Belknap Press, 2002), 213–14.

CHAPTER 3: IMMUNITY IN HYPERDRIVE

53 **More than half of the several dozen:** Frauke Muecksch et al., "Increased Memory B Cell Potency and Breadth after a SARS-CoV-2 mRNA Boost," *Nature* 607, no. 7917 (2022): 128–34, doi.org/10.1038/s41586-022-04778-y.

54 **"Boosting begets breadth":** Duane R. Wesemann, "Omicron's Message on

Vaccines: Boosting Begets Breadth," *Cell* 185, no. 3 (2022): 411–13, doi.org
/10.1016/j.cell.2022.01.006.

54 **Administering doses too close together:** Bruce Y. Lee, "How Long After Having Covid-19 Should You Wait to Get the Booster Vaccine?," *Forbes*, October 7, 2022, forbes.com/sites/brucelee/2022/10/07/how-long-after-having-covid -19-should-you-wait-to-get-the-booster-vaccine.

59 **fifteen hundred different species of microbes:** "Microbiology by Numbers," *Nature Reviews Microbiology* 9, no. 9 (2011): 628, doi.org/10.1038/nrmicro 2644.

60 **He dubbed these factors *Antikörper*:** J. Lindenmann, "Origin of the Terms 'Antibody' and 'Antigen,'" *Scandinavian Journal of Immunology* 19, no. 4 (1984): 281–85, doi.org/10.1111/j.1365-3083.1984.tb00931.x.

60 **Roux suggested that immunity was acquired:** Bartlomiej Swiatczak, "Evolution Within the Body: The Rise and Fall of Somatic Darwinism in the Late Nineteenth Century," *History and Philosophy of the Life Sciences* 45, no. 1 (2023), doi.org/10.1007/s40656-023-00566-7.

60 **In his 1955 paper:** Niels K. Jerne, "The Natural-Selection Theory of Antibody Formation," *Proceedings of the National Academy of Sciences* 41, no. 11 (1955): 849–57, doi.org/10.1073/pnas.41.11.849.

61 **his theory of "clonal selection":** F. M. Burnet, "A Modification of Jerne's Theory of Antibody Production Using the Concept of Clonal Selection," *CA: A Cancer Journal for Clinicians* 26, no. 2 (1976): 119–21, doi.org/10.3322/canj clin.26.2.119.

61 **a theory similar to Burnet's:** Warwick Anderson and Ian R. Mackay, *Intolerant Bodies: A Short History of Autoimmunity* (Johns Hopkins University Press, 2014), 87.

61 **It simply didn't seem possible:** S. Tonegawa, "Somatic Generation of Immune Diversity," *Bioscience Reports* 8, no. 1 (1988): 3–26, doi.org/10.1007/BF01 128968.

62 **the theoretical potential to make a *quintillion*:** Bryan Briney et al., "Commonality Despite Exceptional Diversity in the Baseline Human Antibody Repertoire," *Nature* 566, no. 7744 (2019): 393–97, doi.org/10.1038/s41586 -019-0879-y.

63 **In 1909, Smith showed:** Theobald Smith, "Active Immunity Produced by So Called Balanced or Neutral Mixture of Diphtheria Toxin and Antitoxin," *Journal of Experimental Medicine* 11, no. 2 (1909): 241–56, doi.org/10.1084 /jem.11.2.241.

63 **had co-led an animal study:** Carolyn S. Pincus and Victor Nussenzweig, "Passive Antibody May Simultaneously Suppress and Stimulate Antibody Formation Against Different Portions of a Protein Molecule," *Nature* 222, no. 5193 (1969): 594–96, doi.org/10.1038/222594a0.

64 **possessed a more diverse repertoire:** Dennis Schaefer-Babajew et al., "Antibody Feedback Regulates Immune Memory After SARS-CoV-2 mRNA Vaccination," *Nature* 613, no. 7945 (2023): 735–42, doi.org/10.1038/s41586-022 -05609-w.

66 **"forbidden clone hypothesis":** Maha Alriyami and Constantin Polychronakos, "Somatic Mutations and Autoimmunity," *Cells* 10, no. 8 (2021): 2056, doi.org/10.3390/cells10082056.

66 **approximately 40 percent of early immature B cells:** Barton F. Haynes et al., "Strategies for HIV-1 Vaccines That Induce Broadly Neutralizing Antibodies," *Nature Reviews Immunology* 23, no. 3 (2023): 142–58, doi.org/10.1038 /s41577-022-00753-w.

66 **5 to 10 percent of the general population:** "Autoimmune Disorders Found to Affect Around One in Ten People," University of Oxford press release, May 6, 2023, ox.ac.uk/news/2023-05-06-autoimmune-disorders-found-af fect-around-one-ten-people.

67 **did indeed escape this surveillance:** Christian T. Mayer et al., "An Apoptosis-Dependent Checkpoint for Autoimmunity in Memory B and Plasma Cells," *Proceedings of the National Academy of Sciences of the United States of America* 117, no. 40 (2020): 24957–63, doi.org/10.1073/pnas.2015372117.

67 **mutate to become cancerous:** Alriyami and Polychronakos, "Somatic Mutations and Autoimmunity."

68 **can sometimes be a culprit:** Anne Durandy et al., "Potential Roles of Activation-Induced Cytidine Deaminase in Promotion or Prevention of Autoimmunity in Humans," *Autoimmunity* 46, no. 2 (2013): 148–56, doi.org /10.3109/08916934.2012.750299.

68 **they had less kidney damage:** Ahmad Zaheen and Alberto Martin, "Activation-Induced Cytidine Deaminase and Aberrant Germinal Center Selection in the Development of Humoral Autoimmunities," *American Journal of Pathology* 178, no. 2 (2011): 462–71, doi.org/10.1016/j.ajpath.2010.09.044.

68 **prone to a gamut of autoimmune illnesses:** Zaheen and Martin, "Activation-Induced Cytidine Deaminase."

71 **The discovery of CH103:** Hua-Xin Liao et al., "Co-Evolution of a Broadly Neutralizing HIV-1 Antibody and Founder Virus," *Nature* 496, no. 7446 (2013): 469–76, doi.org/10.1038/nature12053.

74 **able to work against 35 percent:** Wilton B. Williams et al., "Vaccine Induction of Heterologous HIV-1-Neutralizing Antibody B Cell Lineages in Humans," *Cell* 187, no. 12 (2024): 2919–2934.E20, doi.org/10.1016/j.cell .2024.04.033.

75 **the "raw materials of evolution":** Theodosius Dobzhansky, "The Raw Materials of Evolution," *Scientific Monthly* 46, no. 5 (1938): 445–49.

CHAPTER 4: ATTACK OF THE CLONES

77 **"The dream of every cell"**: Jacques Monod, *Chance and Necessity: An Essay on the Natural Philosophy of Modern Biology* (Alfred A. Knopf, 1971), 20.

78 **All the way back in the late 1600s**: Peter Voswinckel, *Der Schwarze Urin—Vom Schrecknis Zum Laborparameter* (Blackwell Wissenschafts-Verlag, 1993), 68.

78 **a leatherworker in London**: William H. Crosby, "Historical Review: Paroxysmal Nocturnal Hemoglobinuria," *Blood* 6, no. 3 (1951): 270–84, doi .org/10.1182/blood.v6.3.270.270; William W. Gull, "A Case of Intermittent Haematuria, with Remarks," *Guy's Hospital Reports* 12, no. 5 (1866): 381–92, babel.hathitrust.org/cgi/pt?id=uc1.b4490373&seq=417.

78 **one in five hundred thousand people**: John Vivian Dacie, "Paroxysmal Nocturnal Hæmoglobinuria," *Proceedings of the Royal Society of Medicine* 56, no. 7 (1963): 587–96, doi.org/10.1177/003591576305600723.

79 **a theory articulated in 1963**: Dacie, "Paroxysmal Nocturnal Hæmoglobinuria."

80 **appeared in a research journal in 1970**: S. B. Oni et al., "Paroxysmal Nocturnal Hemoglobinuria: Evidence for Monoclonal Origin of Abnormal Red Cells," *Blood* 36, no. 2 (1970): 145–52, doi.org/10.1182/blood.v36.2.145.145.

81 **"What Darwin could not anticipate"**: Lucio Luzzatto, "A Journey from Blood Cells to Genes and Back," *Annual Review of Genomics and Human Genetics* 24 (August 2023): 1–33, doi.org/10.1146/annurev-genom-101022-105018.

82 **they saw a strange trend**: Patricia A. Jacobs, W. M. Court Brown, and Richard Doll, "Distribution of Human Chromosome Counts in Relation to Age," *Nature* 191, no. 4794 (1961): 1178–80, doi.org/10.1038/1911178a0.

82 **Two years later another study**: Patricia A. Jacobs et al., "Change of Human Chromosome Count Distributions with Age: Evidence for a Sex Difference," *Nature* 197, no. 4872 (1963): 1080–81, doi.org/10.1038/1971080a0.

82 **loss of Y was present in the patients**: R. I. Holmes et al., "Loss of the Y Chromosome in Acute Myelogenous Leukemia: A Report of 13 Patients," *Cancer Genetics and Cytogenetics* 17, no. 3 (1985): 269–78, doi.org/10.1016/0165 -4608(85)90018-4.

83 **uncovered a twist**: Deborah J. Thompson et al., "Genetic Predisposition to Mosaic Y Chromosome Loss in Blood," *Nature* 575, no. 7784 (2019): 652–57, doi.org/10.1038/s41586-019-1765-3.

83 **associated with a myriad of problems**: Jan Vijg and Xiao Dong, "Pathogenic Mechanisms of Somatic Mutation and Genome Mosaicism in Aging," *Cell* 182, no. 1 (2020): 12–23, doi.org/10.1016/j.cell.2020.06.024; Bela Barros et al., "Loss of Chromosome Y and Its Potential Applications as Biomarker in Health and Forensic Sciences," *Cytogenetic and Genome Research* 160, no. 5 (2020): 225–37, doi.org/10.1159/000508564.

84 **the most common mutation:** Bozena Bruhn-Olszewska et al., "The Effects of Loss of Y Chromosome on Male Health," *Nature Reviews Genetics* 26, no. 5 (2025): 320–35, doi.org/10.1038/s41576-024-00805-y.

84 **in almost 90 percent:** Sarah Zhang, "The Disappearing Y Chromosome," *Atlantic*, December 6, 2019, theatlantic.com/science/archive/2019/12/men -lose-y-chromosomes-cells-they-age/603013/; Lars A. Forsberg et al., "Mosaic Loss of Chromosome Y in Peripheral Blood Is Associated with Shorter Survival and Higher Risk of Cancer," *Nature Genetics* 46, no. 6 (2014): 624–28, doi.org/10.1038/ng.2966.

84 **"highly significant" relationship:** L. M. Russell et al., "X Chromosome Loss and Ageing," *Cytogenetic and Genome Research* 116, no. 3 (2007): 181–85, doi .org/10.1159/000098184.

85 **a paper that grabbed his attention:** Lambert Busque et al., "Recurrent Somatic *TET2* Mutations in Normal Elderly Individuals with Clonal Hematopoiesis," *Nature Genetics* 44, no. 11 (2012): 1179–81, doi.org/10.1038/ng.2413.

86 **a father-and-son research duo:** E. P. Benditt and J. M. Benditt, "Evidence for a Monoclonal Origin of Human Atherosclerotic Plaques," *Proceedings of the National Academy of Sciences of the United States of America* 70, no. 6 (1973): 1753–56, doi.org/10.1073/pnas.70.6.1753.

87 **appeared in *The New England Journal of Medicine*:** Siddhartha Jaiswal et al., "Age-Related Clonal Hematopoiesis Associated with Adverse Outcomes," *New England Journal of Medicine* 371, no. 26 (2014): 2488–98, doi.org/10.1056 /nejmoa1408617.

87 **around 10 to 20 percent:** David P. Steensma and Kelly L. Bolton, "What to Tell Your Patient with Clonal Hematopoiesis and Why: Insights from 2 Specialized Clinics," *Blood* 136, no. 14 (2020): 1623–31, doi.org/10.1182/blood .2019004291.

88 **obesity might exacerbate CHIP:** Santhosh Kumar Pasupuleti et al., "Obesity-Induced Inflammation Exacerbates Clonal Hematopoiesis," *Journal of Clinical Investigation* 133, no. 11 (2023), doi.org/10.1172/jci163968.

88 **Even fragmented sleep:** Cameron S. McAlpine et al., "Sleep Exerts Lasting Effects on Hematopoietic Stem Cell Function and Diversity," *Journal of Experimental Medicine* 219, no. 11 (2022), doi.org/10.1084/jem.20220081.

88 **mutant blood cells in the astronauts' samples:** Agnieszka Brojakowska et al., "Retrospective Analysis of Somatic Mutations and Clonal Hematopoiesis in Astronauts," *Communications Biology* 5, no. 1 (2022): 828, doi.org/10.1038 /s42003-022-03777-z.

88 **people who receive blood:** Sarah S. Burns and Reuben Kapur, "Clonal Hematopoiesis of Indeterminate Potential as a Novel Risk Factor for Donor-Derived Leukemia," *Stem Cell Reports* 15, no. 2 (2020): 279–91, doi.org/10.1016 /j.stemcr.2020.07.008.

89 **twenty-five trillion red blood cells:** Angelo D'Alessandro, "Editorial: Rising Stars in Red Blood Cell Physiology: 2022," *Frontiers in Physiology* 13 (September 2022): 1020144. doi.org/10.3389/fphys.2022.1020144; Ian A. Hatton et al., "The Human Cell Count and Size Distribution," *Proceedings of the National Academy of Sciences of the United States of America* 120, no. 39 (2023): e2303077120, doi.org/10.1073/pnas.2303077120.

89 **2.4 million new red blood cells:** "Blood," Texas Heart Institute, accessed October 1, 2020, texasheart.org/heart-health/heart-information-center/topics/blood-cells.

89 **"This sets the stage":** José J. Fuster and Kenneth Walsh, "Somatic Mutations and Clonal Hematopoiesis: Unexpected Potential New Drivers of Age-Related Cardiovascular Disease," *Circulation Research* 122, no. 3 (2018): 523–32, doi.org/10.1161/CIRCRESAHA.117.312115.

89 **"apply not only to individuals":** Sigurgeir Olafsson and Carl A. Anderson, "Somatic Mutations Provide Important and Unique Insights into the Biology of Complex Diseases," *Trends in Genetics* 37, no. 10 (2021): 872–81, doi.org/10.1016/j.tig.2021.06.012.

90 **encounter a bacterial toxin:** Siddhartha Jaiswal et al., "Clonal Hematopoiesis and Risk of Atherosclerotic Cardiovascular Disease," *New England Journal of Medicine* 377, no. 2 (2017): 111–21, doi.org/10.1056/NEJMoa1701719.

90 **CHIP and heart arrhythmias:** Art Schuermans et al., "Clonal Haematopoiesis of Indeterminate Potential Predicts Incident Cardiac Arrhythmias," *European Heart Journal* 45, no. 10 (2024): 791–805, doi.org/10.1093/eurheartj/ehad670.

93 **which they named "the VEXAS syndrome":** David B. Beck et al., "Somatic Mutations in *UBA1* and Severe Adult-Onset Autoinflammatory Disease," *New England Journal of Medicine* 383, no. 27 (2020): 2628–38, doi.org/10.1056/nejmoa2026834.

93 **twenty thousand or thirty thousand individuals:** David B. Beck et al., "Estimated Prevalence and Clinical Manifestations of *UBA1* Variants Associated with VEXAS Syndrome in a Clinical Population," *JAMA* 329, no. 4 (2023): 318–24, doi.org/10.1001/jama.2022.24836.

97 **slower growth of mutant clones:** Joshua S. Weinstock et al., "Aberrant Activation of TCL1A Promotes Stem Cell Expansion in Clonal Haematopoiesis," *Nature* 616, no. 7958 (2023): 755–63, doi.org/10.1038/s41586-023-05806-1.

99 **his PNH symptoms disappeared:** Daria V. Babushok, "When Does a PNH Clone Have Clinical Significance?," *Hematology: American Society of Hematology Education Program* 2021, no. 1 (2021): 143–52, doi.org/10.1182/hematology.2021000245.

99 ***all* of them had PNH clones:** David J. Araten et al., "Clonal Populations of Hematopoietic Cells with Paroxysmal Nocturnal Hemoglobinuria Genotype

and Phenotype Are Present in Normal Individuals," *Proceedings of the National Academy of Sciences of the United States of America* 96, no. 9 (1999): 5209–14, doi.org/10.1073/pnas.96.9.5209.

101 **one out of six had loss of Y:** Saskia Haitjema et al., "Loss of Y Chromosome in Blood Is Associated with Major Cardiovascular Events During Follow-Up in Men After Carotid Endarterectomy," *Circulation: Cardiovascular Genetics* 10, no. 4 (2017): e001544, doi.org/10.1161/CIRCGENETICS.116.001544.

101 **engineered male mice:** Soichi Sano et al., "Hematopoietic Loss of Y Chromosome Leads to Cardiac Fibrosis and Heart Failure Mortality," *Science* 377, no. 6603 (2022): 292–97, doi.org/10.1126/science.abn3100.

101 **"two sides of the same coin":** Viktor Ljungström et al., "Loss of Y and Clonal Hematopoiesis in Blood—Two Sides of the Same Coin?," *Leukemia* 36, no. 3 (2022): 889–91, doi.org/10.1038/s41375-021-01456-2.

101 **was linked to mutations in *TET2*:** Ahmed A. Z. Dawoud et al., "Age-Related Loss of Chromosome Y Is Associated with Levels of Sex Hormone Binding Globulin and Clonal Hematopoiesis Defined by *TET2*, *TP53*, and *CBL* Mutations," *Science Advances* 9, no. 16 (2023), doi.org/10.1126/sciadv.ade9746.

101 **a reduced risk of loss of Y:** Weinstock et al., "Aberrant Activation of TCL1A."

CHAPTER 5: THE HEART OF THE MATTER

106 **three out of the thirty-six blood cells:** Carl Zimmer, *She Has Her Mother's Laugh: The Powers, Perversions, and Potential of Heredity* (Dutton, 2019), 364.

106 **Astrea had mosaic blood:** Jennie Dusheck, "Girl's Deadly Arrhythmia Linked to Mosaic of Mutant Cells," Stanford Medicine News Center, September 26, 2016, med.stanford.edu/news/all-news/2016/09/newborns-deadly-heart-arrhy thmia-caused-by-mosaic-of-mutant-cells.html.

106 **named Alfred Blaschko, who shared:** Alfred Blaschko, *Die Nervenverteilung in der Haut in ihrer Beziehung zu den Erkrankungen der Haut* (W. Braumèuller, 1901); Hui Jin et al., "Old Lines Tell New Tales: Blaschko Linear Lupus Erythematosis," *Autoimmunity Reviews* 15, no. 4 (2016): 291–306, doi.org/10 .1016/j.autrev.2015.11.014.

107 **Zlotnikov documented a case:** Rudolf Happle and Eckart Haneke, "More About Zlotnikov, the Man Who Explained Blaschko's Lines to Be a Mosaic," *Acta Dermato-Venereologica* 101, no. 8 (2021): adv00521, doi.org/10.2340 /00015555-3880.

107 **rediscovered the lines of Blaschko:** Rudolf Happle, "An Early Description of a 'Human Mosaic' Involving the Skin: A Story from 1945," *Acta Dermato-Venereologica* 100, no. 7 (2020): adv00090-139, doi.org/10.2340/00015555 -3426.

108 **able to confirm the theory:** Samara Silva Kouzak, Marcela Sena Teixeira

Mendes, and Izelda Maria Carvalho Costa, "Cutaneous Mosaicisms: Concepts, Patterns and Classifications," *Anais Brasileiros de Dermatologia* 88, no. 4 (2013): 507–17, doi.org/10.1590/abd1806-4841.20132015.

109 **one of whom had hemimegalencephaly:** Noriko Salamon et al., "Contralateral Hemimicrencephaly and Clinical-Pathological Correlations in Children with Hemimegalencephaly," *Brain: A Journal of Neurology* 129, no. 2 (2006): 352–65, doi.org/10.1093/brain/awh681.

110 **genes that control the growth of neurons:** Carl Zimmer, "DNA Double Take," Science, *New York Times*, September 16, 2013, nytimes.com/2013/09/17/science/dna-double-take.html; Annapurna Poduri et al., "Somatic Activation of AKT3 Causes Hemispheric Developmental Brain Malformations," *Neuron* 74, no. 1 (2012): 41–48, doi.org/10.1016/j.neuron.2012.03.010.

111 **Gleeson and his collaborators published:** Jeong Ho Lee et al., "De Novo Somatic Mutations in Components of the PI3K-AKT3-mTOR Pathway Cause Hemimegalencephaly," *Nature Genetics* 44, no. 8 (2012): 941–45, doi.org/10.1038/ng.2329.

111 **DNA errors in this pathway:** Jean-Baptiste Rivière et al., "De Novo Germline and Postzygotic Mutations in AKT3, PIK3R2 and PIK3CA Cause a Spectrum of Related Megalencephaly Syndromes," *Nature Genetics* 44, no. 8 (2012): 934–40, doi.org/10.1038/ng.2331.

111 **focal cortical dysplasia:** Sang Min Park et al., "Brain Somatic Mutations in MTOR Disrupt Neuronal Ciliogenesis, Leading to Focal Cortical Dyslamination," *Neuron* 99, no. 1 (2018): 83–97.E7, doi.org/10.1016/j.neuron.2018.05.039.

112 **Drugs called mTOR inhibitors:** Alissa M. D'Gama and Christopher A. Walsh, "Somatic Mosaicism and Neurodevelopmental Disease," *Nature Neuroscience* 21, no. 11 (2018): 1504–14, doi.org/10.1038/s41593-018-0257-3.

112 **had this positive outcome:** Se Hee Kim et al., "Efficacy and Safety of Everolimus for Patients with Focal Cortical Dysplasia Type 2," *Epilepsia Open* 10, no. 1 (2025): 243–57, doi.org/10.1002/epi4.13104.

114 **likely a "gross underestimate":** C. I. Edvard Smith, Peter Bergman, and Daniel W. Hagey, "Estimating the Number of Diseases—the Concept of Rare, Ultra-Rare, and Hyper-Rare," *iScience* 25, no. 8 (2022): 104698, doi.org/10.1016/j.isci.2022.104698.

115 **"Scientists previously assumed":** "NIH Researchers Crack Mystery Behind Rare Bone Disorder," National Institutes of Health, April 11, 2018, nih.gov/news-events/news-releases/nih-researchers-crack-mystery-behind-rare-bone-disorder.

115 *two different* **noninherited mutations:** Alison M. Muir et al., "Double So-

matic Mosaicism in a Child with Dravet Syndrome," *Neurology Genetics* 5, no. 3 (2019), doi.org/10.1212/nxg.0000000000000333.

116 **might happen more often:** Catherine Costa et al., "Mosaicism in Men in Hemophilia: Is It Exceptional? Impact on Genetic Counselling," *Journal of Thrombosis and Haemostasis* 7, no. 2 (2009): 367–69, doi.org/10.1111/j.1538 -7836.2008.03246.x.

116 **uncovered a handful of ALPS patients:** Eliska Holzelova et al., "Autoimmune Lymphoproliferative Syndrome with Somatic Fas Mutations," *New England Journal of Medicine* 351, no. 14 (2004): 1409–18, doi.org/10.1056/NEJMoa 040036.

116 **manifest clinically at an older age:** David T. Teachey, "Somatic ALPS: A FAScinating Condition," *Blood* 115, no. 25 (2010): 5125–26, doi.org/10 .1182/blood-2010-04-278465.

116 **identified since the 2010s:** Ankita Singh et al., "An Updated Review on Phenocopies of Primary Immunodeficiency Diseases," *Genes & Diseases* 7, no. 1 (2019): 12–25, doi.org/10.1016/j.gendis.2019.09.007.

117 **An unexpected case turned up:** John M. Ringman et al., "Mosaicism for Trisomy 21 in a Patient with Young-Onset Dementia: A Case Report and Brief Literature Review," *Archives of Neurology* 65, no. 3 (2008): 412–15, doi.org /10.1001/archneur.65.3.412.

118 **a medical case report of a woman:** Jonathan A. Beck et al., "Somatic and Germline Mosaicism in Sporadic Early-Onset Alzheimer's Disease," *Human Molecular Genetics* 13, no. 12 (2004): 1219–24, doi.org/10.1093/hmg/ddh134.

119 **"help improve our practice":** Jason Liebowitz, "Doctors Thought They Knew What a Genetic Disease Is. They Were Wrong," *Atlantic*, January 2, 2025, theatlantic.com/health/archive/2025/01/genomes-somatic-mutations-illness /681173.

119 **One study in *The New England Journal of Medicine*:** M. S. Anglesio et al., "Cancer-Associated Mutations in Endometriosis Without Cancer," *New England Journal of Medicine* 376, no. 19 (2017): 1835–48, doi.org/10.1056/NEJ Moa1614814.

120 **an overrepresentation of certain mutant clones:** Manako Yamaguchi et al., "Spatiotemporal Dynamics of Clonal Selection and Diversification in Normal Endometrial Epithelium," *Nature Communications* 13, no. 1 (2022): 943, doi .org/10.1038/s41467-022-28568-2.

121 **mathematically retrace the development:** Tim H. H. Coorens et al., "Extensive Phylogenies of Human Development Inferred from Somatic Mutations," *Nature* 597, no. 7876 (2021): 387–92, doi.org/10.1038/s41586-021 -03790-y.

121 **a flurry of mutation:** Freek Manders, Ruben van Boxtel, and Sjors Middelkamp,

"The Dynamics of Somatic Mutagenesis During Life in Humans," *Frontiers in Aging* 2 (December 2021), doi.org/10.3389/fragi.2021.802407.

122 **almost three times the number:** Manders et al., "Dynamics of Somatic Mutagenesis."

122 **five early-stage genetic changes:** Tina Hesman Saey, "Some Identical Twins Don't Have Identical DNA," *Science News*, January 7, 2021, sciencenews.org /article/some-identical-twins-dont-have-identical-dna-genetics.

122 **virus-like genetic elements:** Hameed Khan, Arian Smit, and Stéphane Boissinot, "Molecular Evolution and Tempo of Amplification of Human LINE-1 Retrotransposons Since the Origin of Primates," *Genome Research* 16, no. 1 (2006): 78–87, doi.org/10.1101/gr.4001406; Laura F. Campitelli et al., "Reconstruction of Full-Length LINE-1 Progenitors from Ancestral Genomes," *Genetics* 221, no. 3 (2022): iyac074, doi.org/10.1093/genetics/iyac074.

123 **in the brains of mice:** Alysson R. Muotri et al., "Somatic Mosaicism in Neuronal Precursor Cells Mediated by L1 Retrotransposition," *Nature* 435, no. 7044 (2005): 903–10, doi.org/10.1038/nature03663.

123 **in the brain tissue of our own species:** Nicole G. Coufal et al., "L1 Retrotransposition in Human Neural Progenitor Cells," *Nature* 460, no. 725 (2009): 1127–31, doi.org/10.1038/nature08248.

126 **Finally, the paper was ready:** James Rush Priest et al., "Early Somatic Mosaicism Is a Rare Cause of Long-QT Syndrome," *Proceedings of the National Academy of Sciences* 113, no. 41 (2016): 11555–60, doi.org/10.1073/pnas .1607187113.

126 **other examples of heart disorders:** J. Brett Heimlich and Alexander G. Bick, "Somatic Mutations in Cardiovascular Disease," *Circulation Research* 130, no. 1 (2022): 149–61, doi.org/10.1161/circresaha.121.319809.

CHAPTER 6: WHEN OUR BODIES AUTOCORRECT

129 **a 1981 report:** Thomas G. Whitham and C. N. Slobodchikoff, "Evolution by Individuals, Plant-Herbivore Interactions, and Mosaics of Genetic Variability: The Adaptive Significance of Somatic Mutations in Plants," *Oecologia* 49, no. 3 (1981): 287–92, doi.org/10.1007/bf00347587.

129 **"There's a tree growing":** Bert Cregg, "Understanding Tree Reversions," Michigan State University Extension, June 17, 2011, canr.msu.edu/news/un derstanding_tree_reversions.

130 **"Napoleon" oak tree:** Emanuel Schmid-Siegert et al., "Low Number of Fixed Somatic Mutations in a Long-Lived Oak Tree," *Nature Plants* 3, no. 12 (2017): 926–29, doi.org/10.1038/s41477-017-0066-9.

130 **one hundred thousand genetic differences:** Akiko Satake et al., "Somatic Mutation Rates Scale with Time Not Growth Rate in Long-Lived Tropical

Trees," *eLife* 12 (October 2023), doi.org/10.7554/eLife.88456.2; "Study Suggests Long-Lived Tree Species Play Greater Role in Generating Genetic Diversity," Phys.org, June 6, 2023, phys.org/news/2023-06-long-lived-tree-species -play-greater.html.

131 **a beneficial mutation rescuing a tree:** Penelope B. Edwards et al., "Mosaic Resistance in Plants," *Nature* 347, no. 6292 (1990): 434, doi.org/10.1038/347 434a0.

131 **found ten specific DNA changes:** Amanda Padovan et al., "Differences in Gene Expression Within a Striking Phenotypic Mosaic Eucalyptus Tree That Varies in Susceptibility to Herbivory," *BMC Plant Biology* 13, no. 1 (2013): 29, doi.org/10.1186/1471-2229-13-29.

133 **cells with restored dystrophin:** Eric P. Hoffman et al., "Somatic Reversion/ Suppression of the Mouse *mdx* Phenotype In Vivo," *Journal of the Neurological Sciences* 99, no. 1 (1990): 9–25, doi.org/10.1016/0022-510x(90)90195-s.

133 **until their early teens:** Marina Fanin et al., "Dystrophin-Positive Fibers in Duchenne Dystrophy: Origin and Correlation to Clinical Course," *Muscle & Nerve* 18, no. 10 (1995): 1115–20, doi.org/10.1002/mus.880181007.

133 **reported the unusual case:** Jaya Punetha et al., "Somatic Mosaicism Due to a Reversion Variant Causing Hemi-Atrophy: A Novel Variant of Dystrophinopathy," *European Journal of Human Genetics* 24, no. 10 (2016): 1511–14, doi.org /10.1038/ejhg.2016.22.

135 **published the stunning result:** E. A. Kvittingen et al., "Hereditary Tyrosinemia Type I: Self-Induced Correction of the Fumarylacetoacetase Defect," *Journal of Clinical Investigation* 91, no. 4 (1993): 1816–21, doi.org/10.1172 /jci116393.

135 **In a follow-up study:** E. A. Kvittingen et al., "Self-Induced Correction of the Genetic Defect in Tyrosinemia Type I," *Journal of Clinical Investigation* 94, no. 4 (1994): 1657–61, doi.org/10.1172/jci117509.

136 **Scientists in Texas:** Thomas P. Yang et al., "Spontaneous Reversion of Novel Lesch-Nyhan Mutation by HPRT Gene Rearrangement," *Somatic Cell and Molecular Genetics* 14, no. 3 (1988): 293–303, doi.org/10.1007/bf01534590.

137 **reported the findings in 1994:** Rochelle Hirschhorn et al., "Somatic Mosaicism for a Newly Identified Splice-Site Mutation in a Patient with Adenosine Deaminase-Deficient Immunodeficiency and Spontaneous Clinical Recovery," *American Journal of Human Genetics*, July 1994.

137 **they had cracked the case:** Rochelle Hirschhorn et al., "Spontaneous In Vivo Reversion to Normal of an Inherited Mutation in a Patient with Adenosine Deaminase Deficiency," *Nature Genetics* 13, no. 3 (1996): 290–95, doi.org /10.1038/ng0796-290.

137 **"a provocative molecular tale":** Hagop Youssoufian, "Natural Gene Therapy

and the Darwinian Legacy," *Nature Genetics* 13, no. 3 (1996): 255–56, doi .org/10.1038/ng0796-255.

138 **seemed to be miraculously healing:** Marcel F. Jonkman et al., "Revertant Mosaicism in Epidermolysis Bullosa Caused by Mitotic Gene Conversion," *Cell* 88, no. 4 (1997): 543–51, doi.org/10.1016/s0092-8674(00)81894-2.

138 **Two years later, another team found:** Thomas N. Darling et al., "Revertant Mosaicism: Partial Correction of a Germ-Line Mutation in COL17A1 by a Frame-Restoring Mutation," *Journal of Clinical Investigation* 103, no. 10 (1999): 1371–77, doi.org/10.1172/jci4338.

139 **"natural gene therapy":** Joey E. Lai-Cheong, John A. McGrath, and Jouni Uitto, "Revertant Mosaicism in Skin: Natural Gene Therapy," *Trends in Molecular Medicine* 17, no. 3 (2011): 140–48, doi.org/10.1016/j.molmed.2010 .11.003.

140 **As many as 20 percent of individuals:** Eileen Nicoletti et al., "Mosaicism in Fanconi Anemia: Concise Review and Evaluation of Published Cases with Focus on Clinical Course of Blood Count Normalization," *Annals of Hematology* 99, no. 5 (2020): 913–24, doi.org/10.1007/s00277-020-03954-2.

141 **A 1931 rat experiment:** George M. Higgins and Reuben M. Anderson, "Experimental Pathology of the Liver. I. Restoration of the Liver of the White Rat Following Partial Surgical Removal," *Archives of Pathology*, 12 (1931): 186–202.

143 **Zhu's team published the results:** Min Zhu et al., "Somatic Mutations Increase Hepatic Clonal Fitness and Regeneration in Chronic Liver Disease," *Cell* 177, no. 3 (2019): 608–621.E12, doi.org/10.1016/j.cell.2019.03.026.

143 **in a piece commenting on the study:** Miryam Müller, Stuart J. Forbes, and Thomas G. Bird, "Beneficial Noncancerous Mutations in Liver Disease," *Trends in Genetics* 35, no. 7 (2019): 475–77, doi.org/10.1016/j.tig.2019.05.002.

143 **a paper from another group:** Stanley W. K. Ng et al., "Convergent Somatic Mutations in Metabolism Genes in Chronic Liver Disease," *Nature* 598, no. 7881 (2021): 473–78, doi.org/10.1038/s41586-021-03974-6.

146 **the compound shielded the rodents:** Zixi Wang et al., "Positive Selection of Somatically Mutated Clones Identifies Adaptive Pathways in Metabolic Liver Disease," *Cell* 186, no. 9 (2023): 1968–1984.E20, doi.org/10.1016/j.cell.2023 .03.014.

148 **Grompe and his colleagues published a paper:** Ken Overturf et al., "Hepatocytes Corrected by Gene Therapy Are Selected In Vivo in a Murine Model of Hereditary Tyrosinaemia Type I," *Nature Genetics* 12, no. 3 (1996): 266–73, doi.org/10.1038/ng0396-266.

149 **scientists in New York published data:** Michael Oertel et al., "Cell Competition Leads to a High Level of Normal Liver Reconstitution by Transplanted

Fetal Liver Stem/Progenitor Cells," *Gastroenterology* 130, no. 2 (2006): 507–20, doi.org/10.1053/j.gastro.2005.10.049.

150 **Over the course of the experiment:** Anne Vonada et al., "Complete Correction of Murine Phenylketonuria by Selection-Enhanced Hepatocyte Transplantation," *Hepatology* 79, no. 5 (2024): 1088–97, doi.org/10.1097/hep.00000 00000000631.

152 **she also became the first person:** A. Gostynski et al., "Adhesive Stripping to Remove Epidermis in Junctional Epidermolysis Bullosa for Revertant Cell Therapy," *British Journal of Dermatology* 161, no. 2 (2009): 444–47, doi.org /10.1111/j.1365-2133.2009.09118.x.

CHAPTER 7: THE NEXT GENERATION

153 **fifty-two thousand years ago:** N. Morral et al., "The Origin of the Major Cystic Fibrosis Mutation (Delta F508) in European Populations," *Nature Genetics* 7, no. 2 (1994): 169–75, doi.org/10.1038/ng0694-169.

153 **twenty-two thousand and seven thousand:** Kevin Esoh and Ambroise Wonkam, "Evolutionary History of Sickle-Cell Mutation: Implications for Global Genetic Medicine," *Human Molecular Genetics* 30, no. R1 (2021): R119–28, doi.org/10.1093/hmg/ddab004.

153 **around twelve hundred years ago:** Amos Frisch et al., "Origin and Spread of the 1278insTATC Mutation Causing Tay-Sachs Disease in Ashkenazi Jews: Genetic Drift as a Robust and Parsimonious Hypothesis," *Human Genetics* 114, no. 4 (2004): 366–76, doi.org/10.1007/s00439-003-1072-8.

155 **needles in haystacks:** Laurence Loewe and William G. Hill, "The Population Genetics of Mutations: Good, Bad and Indifferent," *Philosophical Transactions of the Royal Society B: Biological Sciences* 365, no. 1544 (2010): 1153–67, doi .org/10.1098/rstb.2009.0317.

156 **150 de novo genetic changes *per generation*:** David Porubsky et al., "Human De Novo Mutation Rates from a Four-Generation Pedigree Reference," *Nature* 643, no. 8071 (2025): 427–36, doi.org/10.1038/s41586-025-08922-2.

156 **A survey of forty-six thousand mutations:** Mónica Lopes-Marques et al., "Meta-Analysis of 46,000 Germline *De Novo* Mutations Linked to Human Inherited Disease," *Human Genomics* 18, no. 1 (2024): 20, doi.org/10.1186 /s40246-024-00587-8.

158 **men with faulty DNA repair genes:** Joanna Kaplanis et al., "Genetic and Chemotherapeutic Influences on Germline Hypermutation," *Nature* 605, no. 7910 (2022): 503–8, doi.org/10.1038/s41586-022-04712-2.

158 **20 to 150 more new germline mutations:** Kitty Sherwood et al., "Germline De Novo Mutations in Families with Mendelian Cancer Syndromes Caused

by Defects in DNA Repair," *Nature Communications* 14, no. 1 (2023), doi.org
/10.1038/s41467-023-39248-0.

159 **patterns associated with chemotherapy:** Kaplanis et al., "Genetic and Chemotherapeutic Influences."

160 **did not trace back to mothers:** Joris A. Veltman and Han G. Brunner, "De Novo Mutations in Human Genetic Disease," *Nature Reviews Genetics* 13, no. 8 (2012): 565–75, doi.org/10.1038/nrg3241; Katherine A. Wood and Anne Goriely, "The Impact of Paternal Age on New Mutations and Disease in the Next Generation," *Fertility and Sterility* 118, no. 6 (2022): 1001–12, doi.org/10.1016/j.fertnstert.2022.10.017.

160 **associated with a growing list:** Victor Stolzenbach, Dori C. Woods, and Jonathan L. Tilly, "Non-Neutral Clonal Selection and Its Potential Role in Mammalian Germline Stem Cell Dysfunction with Advancing Age," *Frontiers in Cell and Developmental Biology* 10 (August 2022), doi.org/10.3389/fcell.2022.942652.

161 **many more mutations came from fathers:** Raheleh Rahbari et al., "Timing, Rates and Spectra of Human Germline Mutation," *Nature Genetics* 48, no. 2 (2016): 126–33, doi.org/10.1038/ng.3469.

161 **A DNA analysis in Iceland:** Ian Sample, "Fathers Pass on Four Times as Many New Genetic Mutations as Mothers—Study," *Guardian*, September 20, 2017, theguardian.com/science/2017/sep/20/fathers-pass-on-four-times-as-many-new-genetic-mutations-as-mothers-study.

162 **sperm of a twenty-five-year-old man:** Wood and Goriely, "Impact of Paternal Age"; Michael White, "Why We're All Mutants," *Pacific Standard*, November 23, 2015, psmag.com/social-justice/we-are-all-x-men; R. John Aitken, Geoffry N. De Iulliis, and Brett Nixon, "The Sins of Our Forefathers: Paternal Impacts on De Novo Mutation Rate and Development," *Annual Review of Genetics* 54 (2020), annualreviews.org/content/journals/10.1146/annurev-genet-112618-043617.

162 **One study of germline mutations:** Augustine Kong et al., "Rate of *De Novo* Mutations and the Importance of Father's Age to Disease Risk," *Nature* 488, no. 7412 (2012): 471–75, doi.org/10.1038/nature11396.

162 **selfish selection of stem cells:** Stolzenbach et al., "Non-Neutral Clonal Selection."

163 **"While sperm mosaicism":** Martin W. Breuss, Xiaoxu Yang, and Joseph Gleeson, "Sperm Mosaicism: Implications for Genomic Diversity and Disease," *Trends in Genetics* 37, no. 10 (2021): 890–902, doi.org/10.1016/j.tig.2021.05.007.

163 **smoking might increase mutations:** Marc A. Beal, Carole L. Yauk, and Francesco Marchetti, "From Sperm to Offspring: Assessing the Heritable Genetic

Consequences of Paternal Smoking and Potential Public Health Impacts," *Mutation Research/Reviews in Mutation Research* 773 (July 2017): 26–50, doi .org/10.1016/j.mrrev.2017.04.001.

163 **Amish individuals show:** Michael D. Kessler et al., "De Novo Mutations Across 1,465 Diverse Genomes Reveal Mutational Insights and Reductions in the Amish Founder Population," *Proceedings of the National Academy of Sciences of the United States of America* 117, no. 5 (2020): 2560–69, doi.org /10.1073/pnas.1902766117.

163 **possibility of screening embryos:** Martin W. Breuss et al., "Unbiased Mosaic Variant Assessment in Sperm: A Cohort Study to Test Predictability of Transmission," *eLife* 11 (2022): e78459, doi.org/10.7554/eLife.78459.

165 **twenty-nine different cell types:** Luiza Moore et al., "The Mutational Landscape of Human Somatic and Germline Cells," *Nature* 597, no. 7876 (2021): 381–86, doi.org/10.1038/s41586-021-03822-7.

166 **an estimated thirty cell divisions:** Wafaa M. Rashed, Erin L. Marcotte, and Logan G. Spector, "Germline *De Novo* Mutations as a Cause of Childhood Cancer," *JCO Precision Oncology* 6 (July 2022), doi.org/10.1200/po.21.00505.

168 **mitochondrial DNA mistakes:** Barbara Arbeithuber et al., "Advanced Age Increases Frequencies of De Novo Mitochondrial Mutations in Macaque Oocytes and Somatic Tissues," *Proceedings of the National Academy of Sciences* 119, no. 15 (2022), doi.org/10.1073/pnas.2118740119.

168 **mutations in blood and saliva:** Barbara Arbeithuber et al., "Allele Frequency Selection and No Age-Related Increase in Human Oocyte Mitochondrial Mutations," *Science Advances* 11, no. 32 (2025): eadw4954, doi.org/10.1126 /sciadv.adw4954.

168 **eggs have tamped down the process:** Aida Rodríguez-Nuevo et al., "Oocytes Maintain ROS-Free Mitochondrial Metabolism by Suppressing Complex I," *Nature* 607, no. 7920 (2022): 756–61, doi.org/10.1038/s41586-022-04979-5.

169 **mechanisms to correct harm:** Jessica M. Stringer et al., "Oocytes Can Efficiently Repair DNA Double-Strand Breaks to Restore Genetic Integrity and Protect Offspring Health," *Proceedings of the National Academy of Sciences* 117, no. 21 (2020): 11513–22, doi.org/10.1073/pnas.2001124117.

CHAPTER 8: STRAINS IN THE SYSTEM

171 **The proportion varies geographically:** Yi-Chu Chen et al., "Global Prevalence of *Helicobacter pylori* Infection and Incidence of Gastric Cancer Between 1980 and 2022," *Gastroenterology* 166, no. 4 (2024): 605–19, doi.org/10.1053 /j.gastro.2023.12.022.

171 **around sixty thousand years ago:** "Association Between Humans and *H. py-*

lori Originated in Africa," *Nature Clinical Practice Gastroenterology & Hepatology* 4, no. 5 (2007): 242–43, doi.org/10.1038/ncpgasthep0783.

172 **mummy nicknamed Ötzi:** Ewen Callaway, "Famous Ancient Iceman Had Familiar Stomach Infection," *Nature,* January 7, 2016, doi.org/10.1038/nature.2016.19127.

172 **"I have had a bad spell":** Letter from Charles Darwin to Joseph Dalton Hooker on December 5, 1863, Darwin Correspondence Project, darwinproject.ac.uk/letter?docId=letters/DCP-LETT-4353.xml.

172 **who has posited that *H. pylori*:** Barry Marshall, "Darwins Illness Was Helicobacter Pylori," *What I Know and What I Think I Know* (blog), February 13, 2009, barryjmarshall.blogspot.com/2009/02/darwins-illness-was-helicobacter-pylori.html.

173 **"It was desperate":** Pamela Weintraub, "The Doctor Who Drank Infectious Broth, Gave Himself an Ulcer, and Solved a Medical Mystery," *Discover,* April 8, 2010, discovermagazine.com/health/the-doctor-who-drank-infectious-broth-gave-himself-an-ulcer-and-solved-a-medical-mystery.

174 **And the work helped:** Barry J. Marshall et al., "Attempt to Fulfil Koch's Postulates for Pyloric Campylobacter," *Medical Journal of Australia* 142, no. 8 (1985): 436–39, doi.org/10.5694/j.1326-5377.1985.tb113443.x.

175 **forty-eight-year-old Tennessee man:** Dawn A. Israel et al., "*Helicobacter pylori* Genetic Diversity Within the Gastric Niche of a Single Human Host," *Proceedings of the National Academy of Sciences* 98, no. 25 (2001): 14625–30, doi.org/10.1073/pnas.251551698.

176 **The reanalysis revealed:** Laura K. Jackson et al., "*Helicobacter pylori* Diversification During Chronic Infection Within a Single Host Generates Sub-Populations with Distinct Phenotypes," *PLOS Pathogens* 16, no. 12 (2020): e1008686, doi.org/10.1371/journal.ppat.1008686.

176 **were more adept at colonizing:** V. P. O'Brien et al., "*Helicobacter pylori* Chronic Infection Selects for Effective Colonizers of Metaplastic Glands," *mBio* 14, no. 1 (2023), doi.org/10.1128/mbio.03116-22.

177 **biopsies from a Lithuanian teenager:** Dangeruta Kersulyte, Henrikas Chalkauskas, and Douglas E. Berg, "Emergence of Recombinant Strains of *Helicobacter pylori* During Human Infection," *Molecular Microbiology* 31, no. 1 (1999): 31–43, doi.org/10.1046/j.1365-2958.1999.01140.x.

177 **accelerate their evolution by a hundredfold:** Ahmad Shafiee et al., "Recombination and Phenotype Evolution Dynamics of *Helicobacter pylori* in Colonized Hosts," *International Journal of Systematic and Evolutionary Microbiology* 66, no. 7 (2016): 2471–77, doi.org/10.1099/ijsem.0.001072.

178 **an evolutionary arms race:** Xavier Didelot et al., "Within-Host Evolution of Bacterial Pathogens," *Nature Reviews Microbiology* 14, no. 3 (2016): 150–62, doi.org/10.1038/nrmicro.2015.13.

178 **infection is "orders of magnitude"**: Bodo Linz et al., "A Mutation Burst During the Acute Phase of *Helicobacter pylori* Infection in Humans and Rhesus Macaques," *Nature Communications* 5 (June 2014): 4165, doi.org/10.1038 /ncomms5165.

178 **was shown to fuel diversity**: Tami D. Lieberman et al., "Genetic Variation of a Bacterial Pathogen Within Individuals with Cystic Fibrosis Provides a Record of Selective Pressures," *Nature Genetics* 46, no. 1 (2014): 82–87, doi.org /10.1038/ng.2848.

179 **"extensively" resistant to treatment**: Didelot et al., "Within-Host Evolution of Bacterial Pathogens."

180 **thirty-nine trillion microbial cells**: Ed Yong, "You're Probably Not Mostly Microbes," Science, *Atlantic*, January 8, 2016, theatlantic.com/science/archive /2016/01/youre-probably-not-mostly-microbes/423228.

180 **One study of hundreds of infants**: Daisy W. Chen and Nandita R. Garud, "Rapid Evolution and Strain Turnover in the Infant Gut Microbiome," *Genome Research* 32, no. 6 (2022): 1124–36, doi.org/10.1101/gr.276306.121.

180 **Finally, by age two**: Fiona Fouhy et al., "Composition of the Early Intestinal Microbiota," *Gut Microbes* 3, no. 3 (2012): 203–20, doi.org/10.4161/gmic .20169.

180 **"may only be the tip"**: Isabel Gordo, "Evolutionary Change in the Human Gut Microbiome: From a Static to a Dynamic View," *PLOS Biology* 17, no. 2 (2019): e3000126, doi.org/10.1371/journal.pbio.3000126.

181 **sixteen of the microbe's genes mutated**: Shijie Zhao et al., "Adaptive Evolution Within Gut Microbiomes of Healthy People," *Cell Host & Microbe* 25, no. 5 (2019): 656–667.E8, doi.org/10.1016/j.chom.2019.03.007.

181 **"The strains of *B. fragilis*"**: "Meet B. fragilis, a Bacterium That Moves into Your Gut and Evolves to Make Itself at Home," ScienceDaily, April 23, 2019, sciencedaily.com/releases/2019/04/190423113958.htm.

182 **higher transfer rates of antibiotic resistance**: Mathieu Groussin et al., "Elevated Rates of Horizontal Gene Transfer in the Industrialized Human Microbiome," *Cell* 184, no. 8 (2021): 2053–2067.E18, doi.org/10.1016/j.cell .2021.02.052.

182 **over the ten-year study**: C. Menni et al., "Gut Microbiome Diversity and High-Fibre Intake Are Related to Lower Long-Term Weight Gain," *International Journal of Obesity (2005)* 41, no. 7 (2017): 1099–105, doi.org/10.1038 /ijo.2017.66.

183 **A remarkable genetic diversity emerged**: Esteban Domingo et al., "Nucleotide Sequence Heterogeneity of an RNA Phage Population," *Cell* 13, no. 4 (1978): 735–44, doi.org/10.1016/0092-8674(78)90223-4.

183 **still rebounded within two weeks**: Gina Kolata, "New AIDS Findings on

Why Drugs Fail," *New York Times*, January 12, 1995, nytimes.com/1995/01/12/us/new-aids-findings-on-why-drugs-fail.html.

183 **as there are in influenza virus samples:** Bette Korber et al., "Evolutionary and Immunological Implications of Contemporary HIV-1 Variation," *British Medical Bulletin* 58 (2001): 19–42, doi.org/10.1093/bmb/58.1.19; John Coffin and Ronald Swanstrom, "HIV Pathogenesis: Dynamics and Genetics of Viral Populations and Infected Cells," *Cold Spring Harbor Perspectives in Medicine* 3, no. 1 (2013): a012526, doi.org/10.1101/cshperspect.a012526.

183 **in a single day:** Barbara S. Taylor et al., "The Challenge of HIV-1 Subtype Diversity," *New England Journal of Medicine* 358, no. 15 (2008): 1590–602, doi.org/10.1056/NEJMra0706737.

185 **the scientists found other mutations:** Ji Hoon Baang et al., "Prolonged Severe Acute Respiratory Syndrome Coronavirus 2 Replication in an Immunocompromised Patient," *Journal of Infectious Diseases* 223, no. 1 (2021): 23–27, doi.org/10.1093/infdis/jiaa666.

185 **"marked within-host genomic evolution":** Victoria A. Avanzato et al., "Case Study: Prolonged Infectious SARS-CoV-2 Shedding from an Asymptomatic Immunocompromised Individual with Cancer," *Cell* 183, no. 7 (2020): 1901–1912.E9, doi.org/10.1016/j.cell.2020.10.049.

190 **The immune response generated:** Pierre Van Damme et al., "The Safety and Immunogenicity of Two Novel Live Attenuated Monovalent (Serotype 2) Oral Poliovirus Vaccines in Healthy Adults: A Double-Blind, Single-Centre Phase 1 Study," *Lancet* 394, no. 10193 (2019): 148–58, doi.org/10.1016/s0140-6736(19)31279-6.

191 **prone to more kinds of mutations:** Gregory D. Ebel, "Mutation, Selection, and Bottlenecks in Polio Vaccine Reversion," *Cell Host & Microbe* 29, no. 1 (2021): 3–5, doi.org/10.1016/j.chom.2020.12.018; Andrew L. Valesano et al., "The Early Evolution of Oral Poliovirus Vaccine Is Shaped by Strong Positive Selection and Tight Transmission Bottlenecks," *Cell Host & Microbe* 29, no. 1 (2021): 32–43.e4, doi.org/10.1016/j.chom.2020.10.011.

CHAPTER 9: CAN WE STOP MUTATING?

194 **possessing this fountain:** Sam Anderson, "Searching for the Fountain of Youth," *New York Times Magazine*, October 24, 2014, nytimes.com/2014/10/26/magazine/my-search-for-the-fountain-of-youth.html.

196 **now called Werner syndrome:** Otto Werner, "On Cataract in Conjunction with Scleroderma," in *Werner's Syndrome and Human Aging*, ed. Darrell Salk et al. (Springer US, 1985), doi.org/10.1007/978-1-4684-7853-2_1; Junko Oshima, Julia M. Sidorova, and Raymond J. Monnat Jr., "Werner Syndrome:

Clinical Features, Pathogenesis and Potential Therapeutic Interventions," *Ageing Research Reviews* 33 (January 2017): 105–14, doi.org/10.1016/j.arr .2016.03.002.

197 **a multifactorial phenomenon:** Satoshi Oota, "Somatic Mutations—Evolution Within the Individual," *Methods* 176 (April 2020): 91–98, doi.org/10.1016/j .ymeth.2019.11.002.

197 **these survivors face an increased risk:** Jan Vijg, "Somatic Mutations, Genome Mosaicism, Cancer and Aging," *Current Opinion in Genetics & Development* 26 (June 2014): 141–49, doi.org/10.1016/j.gde.2014.04.002.

199 **"hits" to the genome:** Gioacchino Failla, "The Aging Process and Cancerogenesis," *Annals of the New York Academy of Sciences* 71, no. 6 (1958): 1124–40.

199 **"On the Nature of the Aging Process":** Leo Szilard, "On the Nature of the Aging Process," *Proceedings of the National Academy of Sciences of the United States of America* 45, no. 1 (1959): 30–45, doi.org/10.1073/pnas.45.1.30.

199 **A major breakthrough:** K. G. Stevenson and H. J. Curtis, "Chromosomal Aberrations in Irradiated and Nitrogen Mustard-Treated Mice," *Radiation Research* 15, no. 6 (1961): 774, doi.org/10.2307/3571114; J. Vijg and J. A. Gossen, "Somatic Mutations and Cellular Aging," *Comparative Biochemistry and Physiology Part B: Comparative Biochemistry* 104, no. 3 (1993): 429–37, doi.org /10.1016/0305-0491(93)90264-6.

199 **In a follow-up experiment:** Howard Curtis and Cathryn Crowley, "Chromosome Aberrations in Liver Cells in Relation to the Somatic Mutation Theory of Aging," *Radiation Research* 19, no. 2 (1963): 337, doi.org/10.2307/3571455.

200 **animals' liver cells accumulated:** Martijn E. T. Dollé et al., "Rapid Accumulation of Genome Rearrangements in Liver but Not in Brain of Old Mice," *Nature Genetics* 17, no. 4 (1997): 431–34, doi.org/10.1038/ng1297–431.

201 **more than just the liver:** Martijn E. T. Dollé and Jan Vijg, "Genome Dynamics in Aging Mice," *Genome Research* 12, no. 11 (2002): 1732–38, doi.org /10.1101/gr.125502.

201 **is a throwback of sorts:** Brandon Milholland, Yousin Suh, and Jan Vijg, "Mutation and Catastrophe in the Aging Genome," *Experimental Gerontology* 94 (August 2017): 34–40, doi.org/10.1016/j.exger.2017.02.073.

202 **hot spots of mutation:** Zane Koch et al., "Somatic Mutation as an Explanation for Epigenetic Aging," *Nature Aging* 5, no. 4 (2025): 709–19, doi.org /10.1038/s43587-024-00794-x.

204 **3,200 mutations within each crypt:** Alex Cagan et al., "Somatic Mutation Rates Scale with Lifespan Across Mammals," *Nature* 604, no. 7906 (2022): 517–24, doi.org/10.1038/s41586-022-04618-z.

206 **researchers called "hypermutated":** Taejeong Bae et al., "Analysis of Somatic

Mutations in 131 Human Brains Reveals Aging-Associated Hypermutability," *Science* 377, no. 6605 (2022): 511–17, doi.org/10.1126/science.abm6222.

207 **its acronym, MiCE:** August Yue Huang et al., "Enrichment of Somatic Cancer Driver Mutations in Alzheimer's Disease Microglia and Its Association with Neuroinflammation," *Alzheimer's & Dementia* 20, no. S1 (2024): e084786, doi.org/10.1002/alz.084786.

207 **neurons from people with Alzheimer's:** Michael B. Miller et al., "Somatic Genomic Changes in Single Alzheimer's Disease Neurons," *Nature* 604, no. 7907 (2022): 714–22, doi.org/10.1038/s41586-022-04640-1.

209 **each cell in the body:** Jan H. J. Hoeijmakers, "DNA Damage, Aging, and Cancer," *New England Journal of Medicine* 361, no. 15 (2009): 1475–85, doi.org/10.1056/NEJMra0804615; Tomas Lindahl, "Instability and Decay of the Primary Structure of DNA," *Nature* 362, no. 6422 (1993): 709–15, doi.org/10.1038/362709a0.

210 **researchers successfully pinpointed:** Ajoy C. Karikkineth et al., "Cockayne Syndrome: Clinical Features, Model Systems and Pathways," *Ageing Research Reviews* 33 (January 2017): 3–17, doi.org/10.1016/j.arr.2016.08.002.

210 **centenarians from Italy:** Paolo Garagnani et al., "Whole-Genome Sequencing Analysis of Semi-Supercentenarians," *eLife* 10 (May 2021): e57849, doi.org/10.7554/eLife.57849.

211 **"connection between DNA repair and aging":** Marisa Venegas, "Long Island Interview: Richard B. Setlow; Genetic Repair: The Role of DNA in Daily Life," *New York Times*, March 5, 1989, nytimes.com/1989/03/05/nyregion/long-island-interview-richard-b-setlow-genetic-repair-the-role-of.html.

214 **cells are exposed to cold:** Hiroyuki Nishiyama et al., "Cloning and Characterization of Human *CIRP* (Cold-Inducible RNA-Binding Protein) cDNA and Chromosomal Assignment of the Gene," *Gene* 204, nos. 1–2 (1997): 115–20, doi.org/10.1016/s0378-1119(97)00530-1.

214 **The group posted their results:** Denis Firsanov et al., "DNA Repair and Anti-Cancer Mechanisms in the Long-Lived Bowhead Whale," preprint, bioRxiv, May 8, 2023, doi.org/10.1101/2023.05.07.539748.

215 **"By treating ageing":** "Investors," Genflow Bioscience, accessed May 11, 2024, genflowbio.com/investors.

216 **a rare variant of the *SIRT6* gene:** Matthew Simon et al., "A Rare Human Centenarian Variant of SIRT6 Enhances Genome Stability and Interaction with Lamin A," *EMBO Journal* 41, no. 21 (2022), doi.org/10.15252/embj.2021110393.

217 **"cellular epigenetic rejuvenation":** Jan Vijg et al., "Mitigating Age-Related Somatic Mutation Burden," *Trends in Molecular Medicine* 29, no. 7 (2023): 530–40, doi.org/10.1016/j.molmed.2023.04.002.

217 **getting rid of cells:** Vijg et al., "Mitigating Age-Related."

219 **hungry raccoons away at night:** Edith Hope Fine, *Barbara McClintock: Nobel Prize Geneticist* (Enslow, 1998), 63.

220 **Clones in CHIP:** Art Schuermans et al., "Clonal Hematopoiesis and Incident Heart Failure with Preserved Ejection Fraction," *JAMA Network Open* 7, no. 1 (2024): e2353244, doi.org/10.1001/jamanetworkopen.2023.53244.

220 **disorders like Alzheimer's disease:** Hind Bouzid et al., "Clonal Hematopoiesis Is Associated with Protection from Alzheimer's Disease," *Nature Medicine* 29, no. 7 (2023): 1662–70, doi.org/10.1038/s41591-023-02397-2; Trisha K. Wathan et al., "The Impact of Dnmt3a or Tet2 CH on Alzheimer's Disease in a Murine Model," *Blood* 140, Supplement 1 (2022): 979–80, doi.org/10.1182/blood-2022-163719.

221 **stem cells with these kinds of mutations:** "How Does CHIP Raise Risk for Blood Diseases?," Jackson Laboratory News and Insights, August 18, 2020, jax.org/news-and-insights/2020/august/how-does-chip-raise-risk-for-blood-diseases.

221 **such as Parkinson's:** Michael A. Lodato and Christopher A. Walsh, "Genome Aging: Somatic Mutation in the Brain Links Age-Related Decline with Disease and Nominates Pathogenic Mechanisms," *Human Molecular Genetics* 28, no. R2 (2019): R197–206, doi.org/10.1093/hmg/ddz191.

223 **"The genome you are conceived with":** Mitch Leslie, "NIH Project Probes the Human Body's Multitude of Genomes," *Science* 381, no. 6659 (2023): 719–20, doi.org/10.1126/science.add3503.

226 **the development of certain cancers:** Elizabeth E. Crouch et al., "Regulation of AID Expression in the Immune Response," *Journal of Experimental Medicine* 204, no. 5 (2007): 1145–56, doi.org/10.1084/jem.20061952.

226 **mitigate AID activity:** Juan Alvarez-Gonzalez et al., "Small Molecule Inhibitors of Activation-Induced Deaminase Decrease Class Switch Recombination in B Cells," *ACS Pharmacology & Translational Science* 4, no. 3 (2021): 1214–26, doi.org/10.1021/acsptsci.1c00064.

227 **"The cultivating struggle":** Wilhelm Roux, *The Struggle of Parts*, trans. David Haig and Richard Bondi (Harvard University Press, 2024), 88.

228 **"everything constantly changes":** Roux, *The Struggle of Parts*, 86.

228 **"Mutation, of course, involves change":** Elof Axel Carlson, *Mutation: The History of an Idea from Darwin to Genomics* (Cold Spring Harbor Laboratory Press, 2011), 3.

Index

INDEX